SPRINGER
LAB MANUAL

Springer-Verlag Berlin Heidelberg GmbH

R. M. Kamp   T. Choli-Papadopoulou
B. Wittmann-Liebold   (Eds.)

Protein Structure Analysis

Preparation, Characterization,
and Microsequencing

With 64 Figures

 Springer

Professor Dr. ROZA MARIA KAMP
Technische Fachhochschule Berlin
FB3 Chemie- und Biotechnik
Seestraße 64
13347 Berlin
Germany

Ass. Professor Dr. THEODORA CHOLI-PAPADOPOULOU
Aristotle University Thessaloniki
School of Biochemistry
Laboratory of Biochemistry
54006 Thessaloniki
Greece

Professor BRIGITTE WITTMANN-LIEBOLD
Max-Delbrück-Centrum
für Molekulare Medizin
Robert-Rössle-Straße 10
13125 Berlin-Buch
Germany

Library of Cogress Cataloging-in-Pulication Data

Protein structure analysis : preparation, characterization, and
   microsequencing / Roza Maria Kamp, Theodora Choli-Papadoupoulou,
   Brigitte Wittmann-Liebold (eds.).
        p.   cm. -- (Springer lab manual)
   Includes bibliographical references and index.
   ISBN 978-3-642-47765-2   ISBN 978-3-642-59219-5 (eBook)
   DOI 10.1007/978-3-642-59219-5
   1. Proteins--Structure--Laboratory manuals. I. Kamp, Roza Maria,
   1951- . II. Choli-Papadopoulou, Theodora, 1956- III. Wittmann
   -Liebold, B. (Brigitte), 1931-   IV. Series.
   QP551.P697643 1997   547.7'5046--DC20   96-35790

© Springer-Verlag Berlin Heidelberg 1997
Originally published by Springer-Verlag Berlin Heidelberg New York in 1997

The use of general descriptive names, registered names, trademarks, etc. in this publication does
not imply, even in the absence of a specific statement, that such names are exempt from the relevant protective laws and regulations and therefore free for general use.

Product liability: The publisher cannot guarantee the accuracy of any information about dosage
and application thereof contained in this book. In every individual case the user must check such
information by consulting the relevant literature.

Cover design: Design & Production GmbH, Heidelberg
Typesetting: M. Masson-Scheurer, Homburg, Saar
SPIN 10527054 31/3137 5 4 3 2 1 0 - Printed on acid-free papier

Preface

"Protein Structure Analysis – Preparation and Characterization" is a compilation of practical approaches to the structural analysis of proteins and peptides. Here, about 20 authors describe and comment on techniques for sensitive protein purification and analysis. These methods are used worldwide in biochemical and biotechnical research currently being carried out in pharmaceutical and biomedical laboratories or protein sequencing facilities. The chapters have been written by scientists with extensive experience in these fields, and the practical parts are well documented so that the reader should be able to easily reproduce the described techniques. The methods compiled in this book were demonstrated in student courses and in the EMBO Practical Course on "Microsequence Analysis of Proteins" held in Berlin September 10–15, 1995. The topics also derived from a FEBS Workshop, held in Halkidiki, Thessaloniki, Greece, in April, 1995. Most of the authors participated in these courses as lecturers and tutors and made these courses extremely lively and successful. Since polypeptides greatly vary depending on their specific structure and function, strategies for their structural analysis must for the most part be adapted to each individual protein. Therefore, advantages and limitations of the experimental approaches are discussed here critically, so that the reader becomes familiar with problems that might be encountered.

Young scientists who are newcomers to the field of protein chemistry may ask why it is necessary to learn all about purifying and characterizing individual proteins and peptides, processes which often are tricky and tedious. Cloning and sequencing the protein's gene is a commonly used and modern method to deduce the sequence of the amino acid chain; hence it may not be necessary to study the protein directly. We often hear the sentence: "we can do it all using molecular biology techniques." However, recent work on yeast and human genome projects and the subsequent discovery of many intron and exon sequences whose biological functions are unknown make the search for and direct study of the protein itself mandatory. Moreover, the shift, 10–15 years ago, from protein studies to gene analysis left many open questions behind, which could be addressed by direct protein isolation and characterization. Young scientists interested in natural science research are strongly advised to learn as much as possible about protein and peptide structural analysis, so that problems encountered in their research can be approached by investigating the proteins involved.

We would like to present a compilation of commonly used protein analysis techniques, but we cannot include all the techniques currently in use. The variability in protein and peptide structure is enormous, ranging from very small polypeptide chains of just a few amino acids to large ones with more than 1000 residues, from very hydrophilic globular structures to largely hydrophobic membrane proteins which may be elongated. A single cell contains thousands of different proteins and peptides; indeed, in the human body the biomass of polypeptides is approximately 17%. These biopolymers are involved in all essential processes in the living cell, such as transport within the body fluids and transfer of molecules through membranes; signal transduction; transcription of the genetic information and its translation into amino acid sequences at the ribosome; and mediators of the immune response. Proteins (enzymes) and peptides (hormones) control and also regulate the biosynthetic pathways. Needless to say correlation of their structure and function is of great importance in understanding biological mechanisms at the molecular level.

Proteins are complicated structures, they form three-dimensional folds of the linear polypeptide chain in a manner which is specific and unique to each of the various proteins. The specificity is fully imprinted in the amino acid sequence; yet, how folding of the linear sequence into the higher order structure is accomplished is still one of the great puzzles in biochemistry. Firstly, the primary structure (amino acid sequence) has to be established; secondly, the secondary structural elements (α-helices, β-sheets, extended structures, turns or coil) must be correlated with the linear sequence, and thirdly, folding of the protein according to the specific higher order (tertiary) structure has to be studied. Finally, interactions with other polypeptide chains or other molecules (quarternary structure) must be identified. Although X-ray structural analysis of protein crystals and NMR studies on proteins in solution are powerful tools to determine tertiary protein structure, the basis for the correct interpretation of these results definitely is knowledge of the amino acid sequence. Any type of modification, such as phosphorylation, methylation, or acetylation, adds to the character of the protein and alters its properties. Furthermore, many proteins are constituents of complexes composed of various proteins and RNA or DNA; they may be bound to lipids and glycosides. Therefore, while it is important to know the sequence and the tertiary structure, and to identify amino acid modifications and the presence of S-S-bonds, it is also necessary to establish whether ligands, cofactors and/or other molecules are complexed to the protein being studied. Finally, the binding sites and binding forces which lead to formation of the stable and fully active complex have to be evaluated. These are the reasons why it is not simple to determine the structure of a protein and to explain its functional role on the molecular level. Many complicated methods must be applied to study the protein's role(s) in the biological process.

The study of proteins includes the use of physical, physicochemical, chemical and biological methods. Although not all of these methods can be described here, we nonetheless hope that we have assembled a suitable collection of modern and sensitive techniques that are easy to perform for beginners and which will facilitate study of a protein's structure and function. The presented methods are well established and were found by experienced investigators to be useful. This book includes purification methods under native and denaturing conditions for structural and functional studies; it describes methods for generation and separation of complex protein mixtures in minute amounts by one- and two-dimensional polyacrylamide gels; it contains highly sensitive amino acid compositional analysis procedures as well as manually and automatically performed sequencing techniques. In addition, database searches are recommended that allow alignment of two sequences or multi-alignments of many proteins, thus enabling identification of proteins by both partial NH_2-terminal and internal sequences obtained by Edman degradation. MALDI-mass spectrometry, which has turned out to be a highly sensitive and fast tool to characterize proteins and fragments by mass analysis, adds to the state of the art methods of protein characterization offered here.

If any questions remain, the authors and editors of this book will be glad to offer their help and assistance on request.

Berlin, September 1996   ROZA MARIA KAMP
                                THEODORA CHOLI-PAPADOPOULOU
                                BRIGITTE WITTMANN-LIEBOLD

Contents

Part I Separation of Proteins 1

1 Purification of Proteins for Sequencing
D.A. KYRIAKIDIS 3

2 Affinity Chromatography of Proteins
H. WOMBACHER and L. JACOB 11

3 Protein Purification by Aqueous
Two-Phase Systems
H. SCHÜTTE 31

4 Rapid HPLC Separation of Proteins, Peptides,
and Amino Acids Using Short Columns Filled
with Nonporous Silica-Based Particles
F. GODT and R. M. KAMP 49

Part II Separation of Peptides 59

5 Enzymatic and Chemical Cleavages of Proteins
R. KRAFT 61

6 Separation of Peptides Using HPLC
and TLC Fingerprints
T. CHOLI-PAPADOPOULOU
and R. M. KAMP 73

7 Microseparation of In Situ Digested Peptides
A. OTTO 85

8 Isolation of Peptides for Microsequencing
by In-Gel Proteolytic Digestion
U. HELLMAN 97

**Part III Manual and Automated
Microsequencing Methods** 105

9 Automated Microsequencing:
Introduction and Overview
B. WITTMANN-LIEBOLD 107

10 A Manual Method for Protein Sequence Analysis
Using the DABITC/PITC Method
T. CHOLI-PAPADOPOULOU, Y. SKENDROS,
and K. KATSANI 137

11 Solid-Phase Sequencing of Peptides and Proteins
J. SALNIKOW 153

12 Sequence Analysis of the NH$_2$-Terminally Blocked Proteins
Immobilized on PVDF Membranes from Polyacrylamide Gels
H. HIRANO 167

Part IV Electrophoresis and Blotting of Proteins 181

13 Two-Dimensional Electrophoresis
P. JUNGBLUT 183

14 Semi-Dry Blotting onto Hydrophobic Membranes
P. JUNGBLUT 215

Part V Analysis of Amino Acids 229

15 Highly Sensitive Amino Acid Analysis
R. M. KAMP 231

16 Quantitative Analysis
of D- and L-Amino Acids by HPLC
D. VOLLENBROICH and K. KRAUSE 249

**Part VI Identification of Cysteine Residues
and Lipids in Proteins** 257

17 Identification of Cysteine Residues
and Disulfide Bonds in Proteins
T. A. EGOROV 259

18 Lipid Modifications of Proteins
T. G. GIANNAKOUROS 269

**Part VII Mass Spectrometry in Peptide
and Protein Analysis** 277

19 Mass Spectrometry in Peptide
and Protein Sequence Analysis
B. THIEDE 279

20 Carboxy-Terminal Sequencing
by Combining Carboxypeptidase Y and P Digestion
with Matrix-Assisted Laser Desorption/Ionization
Mass Spectrometry
B. THIEDE 295

21 Crystallization of Biological Macromolecules
F. FRANCESCHI 303

Subject Index 313

List of Contributors

THEODORA CHOLI-
PAPADOPOULOU
Aristotle University
of Thessaloniki
School of Chemistry
Laboratory of Biochemistry
54006 Thessaloniki, Greece

TSEZI EGOROV
Engelhardt Institute
of Molecular Biology
Analytical Protein Chemistry
Research Unit
ul. Vavilova
32, Moscow 117984
Russian Federation

FRANÇOIS FRANCESCHI
Max-Planck Institut
für molekulare, Genetik
Ihnestraße 73
14195 Berlin, Germany

THOMAS G. GIANNAKOUROS
Laboratory of Biochemistry
School of Chemistry
Aristotle University
of Thessaloniki
Thessaloniki 54006, Greece

FRANZ GODT
Technische Fachhochschule
Berlin, FB3
(Chemie und Biotechnik)

Seestraße 64
13347 Berlin, Germany

ULF HELLMAN
Ludwig Institute for Cancer
Research, Box 595
75124 Uppsala, Sweden

HISASHI HIRANO
Yokohoma City University
Kihara Institute
for Biological Research
Majoka-cho 641–12
Totsuka-ku, Yokohama 244
Japan

LOTHAR JACOB
E. Merck
Abt. Bio- und
Produktionschromatographie
64271 Darmstadt, Germany

PETER JUNGBLUT
Max-Planck Institut
für Infektionsbiologie, AG
Proteinanalytik
Monbijoustraße 2
10117 Berlin, Germany

ROZA MARIA KAMP
Technische Fachhochschule
Berlin, FB3
(Chemie und Biotechnik)

Seestraße 64
13347 Berlin, Germany

KATERINA KATSANI
Aristotle University
of Thessaloniki
School of Chemistry
Laboratory of Biochemistry
54006 Thessaloniki, Greece

REGINE KRAFT
Max-Delbrück-Centrum
für Molekulare Medizin
(MDC)
Robert-Rössle-Straße 10
13122 Berlin, Germany

KERSTIN KRAUSE
Technische Fachhochschule
Berlin, FB3
(Chemie und Biotechnik)
Seestrasse 64
13347 Berlin, Germany

D.A. KYRIAKIDIS
Aristotle University
of Thessaloniki
School of Chemistry
Laboratory of Biochemistry
54006 Thessaloniki, Greece

JOHANN SALNIKOW
Technische Universität Berlin
Institut für Biochemie
und Molekulare Biologie
Franklinstraße 29
10587 Berlin, Germany

HORST SCHÜTTE
Technische Fachhochschule
Berlin, FB3
(Chemie und Biotechnik)
Seestraße 64
13347 Berlin, Germany

YANNIS SKENDROS
Aristotle University
of Thessaloniki
School of Chemistry
Laboratory of Biochemistry
54006 Thessaloniki, Greece

ALBRECHT OTTO
Max-Delbrück-Centrum
für Molekulare Medizin
(MDC)
Robert-Rössle-Straße 10
13122 Berlin, Germany

BERND THIEDE
MDC Forschungsgruppe
Proteinchemie
Max-Delbrück-Centrum
für Molekulare Medizin
Robert-Rössle-Straße 10
13125 Berlin-Buch, Germany

DIRK VOLLENBROICH
Technische Universität Berlin
Institut für Biochemie
und Molekular Biologie
Franklinstraße 29
10587 Berlin, Germany

B. WITTMANN-LIEBOLD
Max-Delbrück-Centrum
für Molekulare Medizin
Research Group Protein
Chemistry
Robert-Rössle-Straße 10
13125 Berlin-Buch, Germany

HELMUT WOMBACHER
Technische Fachhochschule
Berlin, FB3
(Chemie und Biotechnik)
Studiengang Biotechnologie
Seestraße 64
13347 Berlin, Germany

Part I
Separation of Proteins

Purification of Proteins for Sequencing

D. A. KYRIAKIDIS

1.1
Introduction

During the past decade, biochemical methods of protein purification have become so specialized and sophisticated that it is now difficult for the beginner, whether graduate student or specialist in another field, to follow all the minor but important details which lead to a successful procedure. It should always be kept in mind that there are many alternative ways to purify a particular protein. If a particular procedure is not working another must be tried.

I shall not attempt to discuss all the recent progress in the field, which has been well documented elsewhere (Harris and Angal 1989, 1990; Scopes 1994). Rather, some of the basic criteria for purification of a protein will be cited and some of the recent advances in protein purification will be discussed.

1.2
Criteria for Purification of a Protein

To succesfully purify a protein, a suitable strategy is required that is based on background knowledge and on the following criteria (Harris and Angal 1989):

- What is the best source?

  It is usually worthwhile to spend some time searching for a rich source of the protein. The availability and quantity of the source must be considered as well. In the past, for large quantities of a protein, livers from cows or mice or from animals that could be kept in the laboratory were used. Now most biochemists use cultured cells such as bacteria, yeast or mammalian cells. With the advantages of gene cloning techniques, the desired protein can now be expressed in large quantities thereby making purification a much more interesting procedure.

- What is known about the protein?

If a protein has been previously purified from a different source, much of the information can be applied to the protein from the desired source. The size of the molecule, its localization, the pI, its hydrophobicity, the posttranslational modification(s), etc., likely remain the same. If a protein is an enzyme or receptor a successful strategy can be applied based on the relationships of the protein to the substrate or ligand.

- How pure must the protein be?

The extent of purification usually depends on the final use of the protein. If a protein is prepared for research use it must be very clean, whereas if the protein will be used by industry partial purification is more than sufficient.

- How much of the purified protein is required?

For studies on activity, relatively small amounts of the active protein are required, whereas for structural studies larger amounts of highly pure protein will be needed.

- How should the protein be assayed?

To follow a protein during the purification procedure, a quick, reproducible and sensitive assay must exist. This assay should also be cheap, able to be performed in small volumes and not require very expensive instruments.

- How long should purification take?

The purification procedure should be very quick to minimize activity losses, degradation, etc. Usually protein is lost during the different steps of the purification procedure. Therefore, in order to maximize the yield the number of steps should be minimized.

- What will the final cost be?

Cost and time are much more important for commercially used proteins.

- How can a protein be purified?

There are many ways to purify a particular protein. The old procedures included precipitation of proteins with salt, changing the pH of the extract, adsorption, anion exchange, and gel filtration techniques. The new methods include HPLC and FPLC columns, affinity absorbents, immunoaffinity columns and many other techniques described elsewhere (Harris and Angal 1989, 1990; Scopes 1994). These newer approaches allow even the most difficult protein to be purified. Therefore, before any purification procedure is started all the above critical questions should be addressed.

1.3
Purification of Recombinant Proteins

Purification of proteins from transformed cells has become fashionable these days and almost as common as the purification of proteins from natural sources (Harris and Angal 1990; Scopes 1994). The former procedure includes many steps but the critical points to be considered are:

1. Isolation of the particular gene

2. High level expression of the desired protein (high plasmid number, strong promoters, inducible expression)

3. Suitability of the purification procedure from the recombinant cells

A recombinant protein which is expressed in bacterial host cells is usually either an extracellular protein, a membrane bound protein, an intracellular protein, or in the form of inclusion bodies (Harris and Angal 1990; Scopes 1994). Purification of recombinant proteins offers many advantages since it is possible to modify the level of protein in the initial extract. Moreover, the techniques for separation of unwanted proteins are mainly the same as those used in conventional purification methods.

Unfortunately, recombinant proteins are usually not produced in an active form (Schein 1989). The overexpressed eukaryotic and prokaryotic insoluble proteins, the inclusion bodies, can be extracted using strong chaotropic reagents like 6 M urea or 8 M guanidinium chloride and then correctly refolded during or after removal of the denaturant (Marston 1986; Marston et al. 1988). Also, it is possible to purify inclusion bodies proteins by direct addition of affinity resins, without a denaturing and refolding step (Hoess et al. 1988). Problems with the purification of inclusion bodies protein are: (a) Denaturants (urea or guanidinium) are unpleasant to work with and expensive. They can cause irreversible modification of protein structure that will elude all of the most sophisticated analytical tests except the immune system. (b) Refolding must be usually done in very dilute solutions and the protein reconcentrated (Marston 1986; van Kimmenade et al.

1988). The reconcentrating step is complicated by proteolysis and further precipitation of the protein. (c) Refolding encourages protein isomerization, which leads to precipitation during storage.

A new technique, making fusion proteins, has recently been developed. The advantages of this approach are: (a) improved stability of the protein, (b) improvement of those properties of the protein which allow its isolation from the culture extract, (c) faster isolation and (d) use of the protein by biotechnology industries producing either bulk enzymes or high-purity pharmaceuticals.

When the expression level of a protein is low or the purification process is not suitable, a fusion protein should be prepared (Scopes 1994; Schein 1989). Usually a fusion DNA sequence is connected to the gene encoding protein DNA. To simplify purification, either polyarginine or polyhistidine (in the COOH-terminal), protein A, or glutathione transferase can be used. The purification procedures will therefore involve ion exchange column (with ligand polyArg), immunoaffinity column (with ligand IgG) or substrate affinity (with ligand glutathione) and can be easily applied. Finally, cleavage of the fusion part can be achieved by proteolytic enzymes (e.g., carboxypeptidase B) thus releasing the native protein (Scopes 1994; Sassenfeld 1990; Enfords 1992).

1.4
Purification of Membrane Bound Proteins

The extraction of proteins from membranes presents many difficulties (Harris and Angal 1989, 1990; Scopes 1994). In most processes a detergent is used to solubilize the hydrophobic protein from its membrane structure and then to stabilize the extracted protein. The detergents are normally chosen according to the efficiency of extracting the desired protein; however, some consideration for the next purification step should be made (Scopes 1994). Less detergent in a smaller volume is better. Detergents with high critical micelle concentration such as octyl glucoside may improve the fractionation procedure. The most efficient detergents are the strongly ionic sulfates, such as dodecyl sulfate (SDS), and cationic detergents, such as cetyl-trimethylammonium bromide. Purification in the presence of these detergents is limited mainly to size separation by gel electrophoresis.

Although many membrane proteins must be purified in the presence of detergents, in the end it may be necessary to remove the detergent (Buse et al. 1986). In many cases, this will cause inactivation of the protein, which is not a problem if the protein is to be used for sequencing. Hydrophobic beads for adsorbing detergents are available (Furth 1980).

Membrane proteins can be categorized into two general classes: (1) peripheral or extrinsic membrane proteins and (2) integral or intrinsic membrane proteins.

Peripheral membrane proteins are associated with the membrane surface through interactions with either other proteins or the exposed regions of phospholipids. In these cases, the protein may often be dissociated from the membrane by altering the ionic conditions of the buffer or, in more resilient situations, by inducing a degree of denaturation. It should be emphasized here that peripheral membrane proteins are loosely associated and once released do not require any detergent or other solubilizing agent to remain in the buffer. Therefore these proteins, once they have been extracted and any detergent used has been removed, can be treated like any other soluble protein for purification. In all manipulations of membranes, protease inhibitors should be included, since proteolytic enzymes may be activated by, e.g., the presence of EDTA or -SH reagents. To prevent irreversible protein denaturation, the pH of the buffer should remain between 6 and 8 and all incubations must be carried out at 0 °C.

Integral membrane proteins are in contact with the hydrophobic phase of the bilayer and may be classified into four groups, depending on the proportion of their structure that is in contact with the bilayer. Purification of integral membrane proteins in soluble form will require removal of the polypeptide from its membrane anchor (Harris and Angal 1990; Scopes 1994).

- Group I: anchored into the membrane by one or two transmembrane protein segments.

- Group II: utilizes covalent attachment to fatty acids or lipids.

In these two groups the bilayer simply acts as a structural support or means of localization and organization. Examples of proteins in these groups include many enzymes such as peptidases, esterases and phosphatases. The proteins of these two groups can be liberated from the membrane by the use of protease or phospholipase C or D. Treatment by phospholipase C usually leaves a diacylglycerol moiety in the bilayer, while treatment by phospholipase D liberates inositol-linked proteins leaving behind the phosphatidic acid fraction (Harris and Angal 1990; Penefsky and Tzagoloff 1971). Particular care should be taken when using phospholipases since extensive incubation will result in proteolysis within the entire molecule of the membrane protein.

- Group III: possesses a single membrane-spanning segment. Full biological activity of group III proteins depends not only on the two hydrophilic domains, but also on the transmembrane segment. Proteins in this group include different receptors, growth factors and LDL proteins. The globular portion in the aqueous phase can be freed from the rest of the molecule by proteolytic action. Under these conditions it is still possible for the extracellular domain to retain its ligand-binding activity.

- Group IV: proteins that traverse the bilayer several times, with a major fraction of the total structure within the hydrophobic phase. Examples of this group include transfer proteins, receptors or rhodopsin type proteins. Although proteolysis releases fragments of such proteins into the aqueous phase, solubilization of the bilayer almost always requires organic solvents, detergents or chaotropic agents.

Membrane fractions or subcellular organelles can be separated by a number of different approaches. The most successful of these is centrifugation, normally using density gradients. Other approaches include organic solvents for two-phase partitioning, high-voltage free-flow electrophoresis, immunoaffinity, or gel filtration chromatography (Scopes 1994). Release of a membrane protein in soluble form can be achieved by one or more combinations of the following treatments (Penefsky and Tzagoloff 1971): sonication; metal chelators (EDTA, EGTA at 1–10 mM); mild alkaline conditions (pH 8–11); high ionic strength (1 M NaCl); phospholipase treatment, etc. The biological activity of the membrane protein should not be lost and the protein should stay active in solution in a soluble form. Therefore, development of new detergents or chromatographic resins is needed for successful purification of these classes of hydrophobic proteins.

References

Buse G, Steffens GJ, Steffens GCM, Meinecke K, Hensel S, Reumkens J(1986) Sequence analysis of complex membrane proteins (cytochrome c oxidase). In: B. Wittmann-Liebold et al. (eds) Advanced methods in protein microsequence analysis. Springer, Berlin Heidelberg, New York, pp 340–351

Enfords SO (1992) Control of in vivo proteolysis in the production of recombinant proteins. Trends in Biotechnol 10: 310–315

Furth AJ (1980) Removing unbound detergent from hydrophobic proteins. Anal. Biochem 109, 207–215

Harris ELV, Angal S (1990) Protein purification applications: a practical approach, IRL, Oxford

Harris ELV, Angal S(1989) Protein purification applications: a practical approach. IRL, Oxford

Hoess A, Arthur A K, Wanner G, Fanning E (1988) Recovery of soluble, biologically active recombinant proteins from total bacterial lysates using ion exchange resin. Biotechnology 6: 1214–1217

Marston FLA, Angal S, Lowe PA, Chan M, Hill CR (1988) Scale- up of the recovery and reactivation of recombinant proteins. Biochem Soc Tran. 16: 112–115

Marston FAO (1986) The purification of eukaryotic polypeptides synthesized in *E. coli*. Biochemical J 240: 1–12

Penefsky HS, Tzagoloff A (1971) Extraction of water soluble enzymes and proteins from membranes. Methods in Enzymol 22: 204–219

Sassenfeld H M (1990) Engineering protein for purification. Trends Biotechnol 8: 88–93

Schein CH (1989) Production of soluble recombinant proteins in bacteria. Biotechnology 7:1141–1149

Scopes RK (1994) Protein purification: principles and practice (3rd edn). Springer, Berlin Heidelberg New York, pp 1–380

van Kimmenade A, Bond M W, Schumacher JH, Laquoi C, Kastelein RA (1988) Expression, renaturation and purification of recombinant human interleukin 4 from *E. coli*. Eur J Biochem 173, 109–114

Chapter 2

Affinity Chromatography of Proteins

H. WOMBACHER and L. JACOB

2.1
Background and Short Overview

Specific interactions between molecules are fundamental to all biological processes. In particular the interaction between proteins and specific binding components has been well established, e.g., between enzymes and their substrates, activators and inhibitors, between antibody and antigen, between some proteins and particular regions of DNA/RNA, or between receptors and their respective effector molecules such as hormones and transmitters. The technique of affinity chromatography relies on such biospecific affinity. The binding molecule which is fixed to the chromatographic matrix is called the ligand and the protein which is to be purified is called the counterligand or target protein. Specific and reversible binding to the immobilized ligand takes place via a biologically functional site on the surface of the target protein. Thus, affinity chromatography allows purification of most proteins to near homogeneity by utilizing the specific structure of the protein (for reviews see Scouten 1981; Wilchek et al. 1984).

2.1.1
Affinity Interaction and Ligand Classification

The ligand that is to be immobilized on the solid phase should possess the ability to selectively form a stable complex with the target protein. The stability of the selectively formed complex is critically dependent on how closely the ligand fits into the binding site of the protein. Regarding the specificity of binding, however, it is weak forces that play the decisive role, e.g., van der Waals interactions, in which bond strength is inversely proportional to the distance to the seventh power. For this reason sufficiently strong binding requires that the ligand fit precisely into the binding site. Moreover with regard to the dynamic aspects, precise and strong binding can be additionally supported by conformational changes, a process sometimes also described as induced fit. For practical purposes, in most cases the binding constant (K_a), which is in the range of $10^5-10^8\,M^{-1}$, is suitable for the affinity chromatographic process. Affinities that are too low do not retard enough interacting target protein during the chromatographic run, resulting in too little selectivity. If affinities are too high, then very harsh, drastic and sometimes denaturing conditions are required for elution of the protein. However, strong binding of the target protein to the column does not necessarily mean specific binding. The occurrence of nonspecific, strongly adsorptive binding forces between the target protein and the chromatographic material which overlap or superimpose the specific binding forces must also be taken into account. In this context specific binding is interpreted as an adsorption that refers only to a locally restricted area on the protein surface, i.e., the binding site.

For the purpose of a better overview, ligands are classified as mono-specific or group-specific. Furthermore they are subdivided into low molecular weight or high molecular weight (macromolecular) ligands.

Mono-specific ligands Monospecificity refers to those ligands that bind to a single or very small number of proteins; steroid hormones and enzyme inhibitors are representative of this type of ligand. A well known example is the low molecular weight ligand biotin (K_a of the biotin-avidin/streptavidin complex = $10^{15}\,M^{-1}$). This feature of highly specific and strong binding has been a valuable tool in biochemical methods similar to AFC (see Wilchek and Bayer 1990).

Macromolecuar mono-specific ligands Antibody-antigen binding is the most important example of a macromolecular mono-specific ligand and its target protein. The antibody, as the ligand bound to a matrix, is a suitable tool for purifying soluble proteins, peptides, solubilized membrane proteins, and even viruses and cells. With the advent of modern hybridoma technology, polyclonal antibody preparations can be replaced by monoclonal antibodies (regarding practice and theory see Peters and Baumgarten 1989). Highly specific antibodies can be obtained against any antigenic protein, even if it is present in only very small concentrations with respect to the total immunogen. Suitable screening methods lead to hybridoma cells producing antibodies of the desired specificity and affinity (Tijssen 1985). When immobilized to an appropriate chromatographic material, these antibodies allow purification of the antigen of interest.

Group specificity refers to ligands, including coenzymes, vitamins/cofactors or their analogues, which bind to sites common to several different proteins. Examples of group-specific ligands include ATP and NAD. The list of target proteins includes cofactor-dependent enzymes, such as NAD- or NADP-dependent dehydrogenases.

Group-specific ligands

Lectins and some immunoglobulin-binding proteins are kown as group-specific and macromolecular ligands. Immobilized lectins are highly specific for sugar residues. For example, concanavalin A is used for purification of some glycoproteins by binding to α-D-mannopyranosyl, α-D-glucopyranosyl and sterically related residues. Staphylococcal protein A and streptococcal protein G are well known as specific ligands in the affinity chromatography of antibodies (Table 1).

Macromolecular group-specific ligands

Table 1. Antibody-binding characteristics of protein A, protein G and Fractogel EMD TA

Antibody	Fractogel EMD TA	Protein A	Protein G
Human IgG1	+	+	+
Human IgG2	+	+	+
Human IgG3	+	–	+
Human IgG4	+	+	+
Human IgM	+	+	–
Human IgA	+	+	–
Mouse IgG1	+	Weak	+
Mouse IgG2a	+	+	+
Mouse IgG2b	+	+	+
Mouse IgG3	+	–	+
Rat IgG1	+	Weak	+
Rat IgG2a	+	–	+
Horse IgG	+	–	+
Goat IgG	+	+	+
Chicken IgG	+	–	–
Bovine IgG	+	Weak	+
Rabbit IgG	+	+	+
Sheep IgG	+	–	+
Dog IgG	+	+	–
Pig IgG	+	+	+
Cat IgG	+	+	–
Chicken (yolk) IgY	+	–	–
Recombinant ab: scFv	+	–	–

2.1.2
Pseudobiospecific Affinity Interaction and Artifical Ligands

Special artificial compounds, e.g., anthraquinone-, or azo-dyes, can be successfully applied as ligands due to their affinity for a wide variety of proteins, especially enzymes such as dehydrogenases, kinases, transferases, and reductases. In some cases the so-called biomimetic interaction is thought to be a result of selective binding of the artifical ligand to the dinucleotide-fold domain, maintained throughout evolution in various proteins. Although the interaction with the protein is not yet clearly understood, this type of affinity chromatography is highly popular (Scopes 1987). The term pseudoaffinity chromatography, also known as pseudobiospecific affinity chromatography, describes the specific interaction of an artificial ligand with certain groups of proteins. In this sense, even covalent binding between ligands, e.g., boronic acid derivatives and some glycoproteins, also referred to as covalent chromatography, could be included. In general, pseudoaffinity chromatography refers to all types of adsorption chromatography that are based on biomimetic ligands.

Immobilized metal ion affinity chromatography/thiophilic adsorption chromatography

Immobilized metal ion affinity chromatography (IMAC) (Porath and Olin 1983; for review see Kagedal 1989) and thiophilic adsorption chromatography (TAC) (Hutchens and Porath 1986) are important examples of pseudoaffinity chromatography. For purification of immunoglobulins TAC has practical advantages over affinity chromatography, which uses protein A as the ligand. These advantages and the successful application of both IMAC and TAC are described in Sect. 2.2.

2.1.3
Chromatographic Supports

The basic requirements of a support which is suitable for affinity chromatography (see Mohr and Pommerening 1985; Narayan and Crane 1990) are: (a) a hydrophilic and neutral-acting matrix; (b) good chemical, mechanical, biological and thermal stability; (c) additional functional groups, e.g., $-OH$, $-NH_2$, $-COOH$, and $-CHO$, which allow derivatization by a wide variety of chemical reactions. For protein purification especially, high macroporosity is preferred. Agarose is the most popular matrix, especially in the stabilized cross-linked form. It possesses most of the characteristics of an ideal matrix for affinity chromatography of proteins. The porous nature of the support more or less impedes diffusional mass transfer, which in affinity chromatography of proteins is as important as the kinetic aspects of binding and disso-ciation of the target protein. Synthetic polymers based on poly-acryl, polyvinyl, polystyrene and particular copolymerizations of suitable monomers have led to the development of nearly ideal matrices, with and without reactive groups, for ligand immobili-zation. Besides good chemical and physical stability, the follow-ing matrices also provide a high macroporosity: Eupergit (Fluka, Neu-Ulm, Germany), Spheron (Waters, Milford, MA, USA), TSK (Tosohaas, Philadelphia, PA, USA), and Fraktogel, (Merck, Darmstadt, Germany). Furthermore, a relatively new and inno-vative matrix made of polystyrene-divinylbenzene, which is coated with an hydroxylated polymer and which eliminates the mass transport bottleneck of conventional chromatography, has recently been described (Regnier 1991). The tradename of this material is POROS, available from PerSeptive Biosystems (Cambridge, MA, USA).

2.1.4
Immobilization Procedure

Immobilization of the ligand to the matrix material is carried out in several steps (for details of the methods see Hermanson et al. 1992; Carlson et al. 1989; Dean et al. 1985; Wong 1991).

Chemicals such as BrCN, carbonyldiimidazole, tosyl or tresyl chloride, bisoxirane, chloroformiate esters, divinylsulphone and hydrazine (**Note**: in some cases special safety precautions are needed) are used to derivatize functional groups of the matrix such that electrophilic groups are generated which are conse-quently reactive to nucleophilic groups (e.g., $-OH$, $-NH_2$, $-SH$) of ligands or vice versa. Alternatively, several different preactivated chromatographic supports are commercially available. In par-ticular, synthetic polymers such as Fractogel EMD Epoxy 650 or Fractogel EMD Azlactone 650 (Merck, Germany) can be recom-mended, if along with chemical and physical stability macro-porosity is necessary. The high capacity of reactive oxirane groups, or azlactone groups respectively, is provided by mul-tipoint attachment to linear polymer chains (called tentacles) grafted on the Fraktogel material. Compared to the conventional spacer technique, it offers some advantages with respect to both capacity and flexibility of the coupled ligand and the protein-binding process; in particular, nonspecific interaction between the proteins and the polymer surface can be avoided (see also Sect. 2.2).

Activation of the matrix

Coupling of the ligand A covalent stable linkage is formed mostly by nucleophilic attack of the ligand on the activated matrix material. Subsequently, the ligand is regarded as immobilized. For instance, with an oxirane-activated or azlactone-activated matrix material, very stable ether and, respectively, extraordinarily stable ester binding is formed. By contrast, a BrCN-activated support material forms much less stable carbiminoester binding; moreover, an often undesired, charged imino group arises at the commonly used pH.

Spacer arms If small molecules are used as ligands to bind macromolecules such as proteins, then so-called spacer arms are recommended. Short aliphatic, linear, hydrocarbon chains with two functional groups at each of the ends of the chains are attached to the matrix. The functional group, which remains free and which is called the terminal group, is then used to couple the ligand by means of a suitable chemical method. The main reason for coupling low molecular weight ligands via spacer molecules is that the ligand binding site of a protein is often buried or in a pocket just below the surface. A ligand directly attached to the matrix may not protrude far enough away from the matrix surface to contact the binding site on an approaching protein molecule. Therefore, a very important aspect is not only the absolute number of ligands immobilized to the matrix, but also their accessibility. Spacer molecules between ligands and the matrix (or so-called tentacles, as provided by Fractogel EMD material) are thus very helpful in improving the specific binding of target proteins and, consequently, enhancing the capacity to bind the target protein. A suitable spacer may also hinder nonspecific adsorption of the protein to the surface of the matrix material.

Site-directed immobilization Site-directed immobilization is an important aspect of the coupling process, as demonstrated by the following example: Obviously the binding potential of a bivalent antibody for an antigen can only be realized fully when the antibody molecule is suitably bound, e.g., via the carbohydrate portion, to the matrix, thereby leaving the binding sites free for interaction with antigen. The proper immobilization chemistry which provides covalent attachment away from the antigen binding sites has comprehensively been described by Hermanson et al. (1992). There are other examples in which site-directed immobilization is a prerequisite for successful binding of target protein: cytochrome c oxidase can be purified by affinity chromatography using cytochrome c as the biologically functional binding ligand, only if it is suitably site-directed and immobilized to the chromatographic support (Azzi et al. 1982).

At the matrix, residual active groups have to be inactivated before affinity chromatography can be performed. This can be done
by an excess of an hydrophilic low molecular weight compound
such as ethanolamine; however, the introduction of positive-
($-NH_3^+$) or negative-($-COO^-$) charges, which can create undesired ion-exchange effects and which arise dependent on the pH,
must be avoided. An increase in hydrophobicity by the blocking
reagent should, as far as possible, also be prevented.

Blocking of residual active groups

2.1.5
Chromatographic Run

Although a one-step purification procedure may be tempting,
preliminary steps generally provide certain advantages: (a) removal of particles, cell debris, membrane fragments, etc.; (b)
delipidation of the sample and the removal of the proteases or
their inhibition. Pretreatment of the sample by protein precipitation or ion-exchange chromatography easily removes many
unwanted substances. Speed and recovery are the key factors
during this stage, and, as a rule, a concentration step is not necessary.

Sample preparation

Specific binding of the target protein to the ligand requires optimization of pH, buffer salt composition and ionic strength.
Obviously, pH, which regulates the presence of charged groups
on the target protein and possibly on the ligand, can have a decisive effect on the binding process (as well on the desorption
step). Some general rules are helpful:

Binding step

- A moderate salt concentration (0.1–0.15 M NaCl) stabilizes
  proteins in solution and prevents nonspecific interactions by
  ion-exchange effects.

- Detergents (e.g., Tween-20) can be used as blocking additives
  to reduce nonspecific interactions owing to hydrophobicity of
  the matrix material, the chemistry of immobilization and/or
  properties of the spacer.

- Very high salt concentrations (e.g., ammonium sulfate up to
  2 M) specifically enhance hydrophobic interactions of ligand
  and target protein.

- EDTA (about 10 mM) is used to stabilize -SH groups of proteins which could be oxidized catalytically by heavy metals.

- Optimization of the binding buffer composition is often the
  key to elimination of nonspecific binding in affinity chromatography.

Column operation   Column size is governed by the capacity of the adsorbent and the amount of protein to be purified. As a rule, high capacity allows for short, wide columns. In most cases, disposable plastic mini-columns (commercially available from, e.g., Bio Rad, Pierce, USA) and 1–5 ml of gel can be used. The hydrophobic character (polystyrene and polypropylene material) of the column and of the top and bottom porous disk is eliminated by treatment with a 0.1% Tween-20 solution. However, longer columns are recommended if the target protein is retarded on the column material by a low binding strength ($K_a < 10^4 \, M^{-1}$). In this case separation of the desired protein from the unbound contaminants depends critically on the column length.

Flow rate   A low flow rate is recommended for sharp elution peaks and maximal recovery with minimal dilution of the purified protein. Too high flow rates reduce both the resolution and the capacity, as it often takes a long time until the new equilibrium is reached. On the one hand the dissociation kinetics of the complex can be rate limiting, in particular, when competitive elution with the soluble ligand/analog is performed. Assuming a first order reaction, the dissociation rate constant correlates indirectly to the strength of the binding of the protein (i.e., half-life time = ln 2/k–1). On the other hand the degree of macroporosity of the matrix material can lead to considerable mass transfer restrictions, dependent on the molecular weight and the shape of the target protein. For these reasons it may be sometimes effective to stop the flow (for 0.5–2 h) after applying the competing agent, particularily when the target protein is very tightly bound to the column. In summary, the linear flow rate through the affinity column has to be optimized to find the maximal binding capacity and the best separation. Most affinity columns operate under gravity flow at or near optimal binding levels. HPLC, however, can only be used if the matrix material is sufficiently pressure resistant. Furthermore, macroporous matrix materials are often not sufficiently stable, although some synthetic matrix materials, e.g. Fractogel, have enough physical stability and thus can be used at relatively high pressure (see Sect. 2.1.3).

Washing step   After the adsorption step the column has to be washed with several volumes of the starting buffer to remove all unbound material. Nonspecific binding, which is based on electrostatic interactions on the column, can be reduced by washing with buffers with somewhat increased ionic strength. The washing steps, observed by UV monitoring, are finished when the original baseline is reached.

The correct choice of the elution conditions is as important as **Elution step** determining the optimal binding conditions. An optimal elution will permit complete recovery of the target protein in a minimum volume. Usually the biological activity of the target protein must also to be maintained. The most basic approach involves competitive displacement of bound protein by soluble ligand, or an analog, at a relatively moderate concentration (about 0.2–1 M) and under otherwise mild desorption conditions (e.g., neutral pH). The mode of elution is stepwise, by gradient or pulse elution; however, extraordinarily strong binding, as occurs in the biotin-avidin complex ($K_a = 10^{15}\,M^{-1}$), is resistant to this desorption procedure. In such a case, in addition to elution with soluble ligand at high concentrations, the harsh conditions of 6 M guanidine-HCl and pH 1.5 are needed for dissociation of the complex within a reasonable time. In principle, the affinity constant is greatly influenced by the composition of the solvent. This must be considered regarding the binding step as well as the elution step. Therefore, specific competitive elution can be carried out effectively by a suitable change in the buffer solvent. In fact, in many cases the specific eluting agent can be completely replaced by so-called nonspecific elution. Of course, a compromise has to be made between the harshness of the eluants and the risk of denaturing the target protein. The following list consists of several procedures that gradually increase the elution conditions:

- A change in pH and/or ionic strength: For example, a NaCl salt gradient can be used to elute proteins predominantly bound by specific electrostatic interactions.

- If hydrophobic interactions between ligand and target protein predominate and a change in pH is not sufficiently effective, more drastic methods are needed such as reduction of the polarity of the aqueous solvent by, e.g., ethylene glycol (up to 60%) or DMSO (up to 10%). The disadvantages are the influence of these agents on viscosity and their effects on polysaccharide matrices.

- Chaotropic salts such as KSCN or KI at high concentrations (up to 3 M) and/or denaturants, such as urea or guanidine salts in still moderate concentrations (4–6 M) can be applied. These salts influence the hydrophobic interaction following the so-called Hofmeister series, as discussed by Eriksson (1989).

- Detergents also interact highly effectively with hydrophobic binding sites. They are to be used at concentrations just below the critical micellar point. In this context it should be noted that it is also helpful to include detergents, at low concentrations, during the entire purification process, especially when membrane proteins are involved. Detergents suppress nonspecific adsorption to the walls of the column and to column material and hinder undesired aggregation of proteins.

When the target protein has been eluted by specific or nonspecific means or by a combination of both, it must then be dialysed or gel-filtrated to remove the eluent and to restore the native conformation if necessary.

Regeneration of the column A reequilibration with several column volumes of the starting buffer is sometimes sufficient to regenerate the affinity column, but undefined material is often still bound to the column and has to be removed by harsher treatment. However, there are some limitations due to the stability of the matrix material, the nature of the ligand and the mode of its covalent linkage to the support.

2.2
Special Applications and Experimental Procedure

2.2.1
Immobilized Metal Chelate Affinity Chromatography

Principle This highly efficient method is based on the interaction of a metal ion chelating agent, covalently bound on a chromatographic support, with proteins containing histidine on the surface (Porath et al. 1975; Sulkowski 1985). In the case of Fractogel EMD iminodiacetic acid has been chosen as the metal chelating ligand. This bidentate chelating moiety remains free after immobilization so that various metal ions can be bound on the stationary phase and particular proteins or peptides are bound via the free coordination sites of the metal ion. The iminodiacetic acid is immobilized via tentacles (as described above) which carry the ligands covalently bound in high density thereby guaranteeing a high degree of hydrophilicity, flexibility and capacity. The tentacles are superior to conventional spacer molecules, since the capacity for protein binding is considerably increased. The protein binding capacity for lysozyme, e.g., is about 60 mg/ml of copper-loaded gel. Additionally, due to the highly flexible immobilzed ligands the metal ions are favorably arranged relative to the binding sites on the protein's surface, which results in improved complex formation and proteins that are very tightly bound to the Fractogel EMD chelating material. Of course, the accessibilty of the histidine residues on the surface is just as important for protein binding (Hemdan et al. 1989).

With respect to peptides, the α-amino group can also play a decisive role, retarding peptides even if no histidine residue is present. It has also been reported that other metal binding amino acids such as cysteine and tryptophane can contribute to the binding of proteins (Hansen and Lindberg 1992, 1995).

Owing to advances in molecular biology, histidine residues can be inserted into a protein's sequence. These modified proteins can be regarded as promising candidates for IMAC, alleviating the otherwise laborious purification of some recombinant proteins. If necessary, the fusion part of the protein could be cleaved chemically or by a specific enzyme after purification. **Recombinant proteins**

The binding of proteins to the Fractogel EMD chelating column is commonly performed at a neutral pH due to the pK of the histidine group. However, the actual pK can greatly vary – up to more than one pH unit depending on the neighboring amino acid residues in the protein.Thus a buffer at pH 8 often leads to an improved binding, but some buffers within this pH region are not suitable. For example, a Tris-containing buffer reduces the binding affinity and should therefore only be used if the proteins have very strong interactions with the affinity matrix. In principle, all substances, such as imidazole, histidine or glycine, which compete with the proteins/peptides for binding to the metal ions should be avoided during the adsorption step; however, they are obviously very well suited for the elution step. During all column operations a high ionic strength is recommended due to the possible ionic interactions between proteins and carboxy groups, which might resist binding to complexing metal ions. The presence of 0.5–1 M NaCl is sufficient; sometimes even 0.1 M NaCl is enough. The elution can be performed by displacement of the protein with a competitive molecule and/or by changing the pH. Displacement of the protein from the column takes place when an appropriate concentration of the competitive agent is reached. Mostly a gradient generated with imidazole is used. A decrease of the pH by using as a step gradient or as linear gradient also reduces the affinity of the protein-metal ion complex and the protein can be desorbed. **Chromatographic conditions**

Compared to true biospecific affinity chromatography, pseudobiospecific affinity techniques such as IMAC are easier and faster to perform. Along with the advantages of its wide range of applications and comparatively low cost, there is the compatibility of IMAC with high salt concentrations in the buffer. Since no interference with the binding of proteins or peptides on the column is observed, the pooled fractions from anion or cation exchangers containing salt at high concentrations can directly be applied **Advantages of IMAC** onto a metal chelating affinity column. Thus, in most cases, time-consuming rebuffering or dialysis of the sample before loading onto the column is not necessary. Furthermore, due to the fact that various metal ions can be complexed to the column, various selectivities can be generated with the same sorbent. Generally the strongest binding of proteins is observed using copper ions followed by nickel ions, with zinc or cobalt ions resulting in weaker binding strengths.

2.2.2
Thiophilic Adsorption Chromatography

Principle   The high efficiency of this type of pseudoaffinity chromatography is based on the salt-promoted adsorption of some proteins, especially antibodies, to heteroaliphatic ligands containing sulfone and thioether groups (Porath et al. 1985). Binding of the protein takes place mainly via accessible tryptophane and/or phenylalanine residues, motifs which are observed within conserved regions of antibody molecules. Thus, proteins with so-called thiophilic regions such as immunoglobulins (Belew et al. 1987) and peptides with aromatic amino acid residues are well suited for TAC. Since albumins are not adsorbed on the thiophilic media, effective separation of antibodies from serum is greatly simplified. Fractogel EMD TA is designed for the efficient isolation of antibodies, and all antibodies tested so far bind to Fractogel EMD TA (see Table 1). As recently described (Schulze et al. 1994), an efficient purification of recombinant, single-chain antibody fragments produced in *E. coli* cells was successfully performed by the thiophilic adsorption technique. The isolation and purification of other proteins and peptides carrying thiophilic areas on the surface by this technique have also been reported (Bischoff et al. 1994). With Fractogel EMD TA, the problem of relatively low protein binding capacities has been overcome by using the tentacle principle. As decribed above, the tentacles carry the group-specific ligands in a high density. Thus, the protein binding capacity for γ-globulin amounts to approximately 30 mg/ml gel. The tentacle principle also provides for an appropriate spacing of the functional ligand molecules, which results in minimized nonspecific interactions of the column with the proteins and high protein binding capacity.

Chromatogra-   Since the adsorption of proteins on Fractogel EMD TA is a salt-
phic conditions   promoted process, the influence of the type of salt and its concentration on the binding efficiency has to be considered (see also Sect. 2.1.5). The best results for antibody binding are ob-

tained at ammonium sulfate concentrations between 0.8 M and 1.5 M. The elution step is easily performed by a decreasing salt gradient. However, at higher concentrations of ammonium sulfate the Fractogel EMD TA behaves increasingly like the material used in hydrophobic interaction chromatography (HIC) (Eriksson 1989).

With respect to antibody purification, the great advantage of TAC is the high chemical stability of the ligand, especially compared with other biospecific ligands, e.g., the macromolecular ligands protein A or protein G, which are also often used. Another advantage of the TAC method is that it fits well into the standard protein purification scheme. Following an ammonium sulfate precipitation or an ion exchange step, the salt concentration remaining in the sample only has to be adjusted (e.g., via the conductivity) to that salt concentration necessary for binding of the target protein to the column. Also, it is of interest that antibodies from very different species and various immunoglobulin classes bind to the thiophilic sorbent, whereas protein A and, respectively, protein G columns bind some antibodies only weakly and others not at all (Table 1).

Advantages of TAC

2.2.3
Material and Methods

First, the purification of peptides by IMAC, performed on Fractogel EMD chelate, will be described in detail followed by the purification of antibodies TAC using Fractogel EMD TA. With slight modifications or even without any changes both protocols can be applied to other peptide and/or protein separations.

The columns can be connected to any HPLC or medium pressure system. The pumps should work precisely with constant flow rates. An inert pump system is preferred, because the buffers contain high concentrations of salt. Also the column lifetime may be prolonged if an inert system is used. Stainless steel components may corrode if high salt buffers are used, giving rise to metal oxides that bind to the chromatographic support. This may result in a loss of resolution and recovery and will eventually damage the column. The minimum equipment required is a gradient mixer, an UV detector (280 nm) and a medium pressure pump. Of course, a programmable gradient mixing system is helpful in more exactly defining the conditions. If a peristaltic pump is used, the volumetric flow rate should be calibrated using the buffer with a high salt concentration. The tubing should be as

Equipment

short as possible to reduce undesired dilution. If the target proteins are very labile, all steps should be carried out at 4 °C.

Gel preparation Fractogel EMD Chelate and Fractogel EMD TA gels are supplied as suspensions. First, the storage solution has to be removed from the gel.

1. Shake the storage container with the gel until a homogeneous suspension is obtained.

2. Place the appropriate amount of gel suspension (8–10 ml to pack a 50 x 10 cm column) in a sintered glass funnel. Suck off the supernatant and resuspend the gel three times in a three-fold excess of equilibration buffer, followed by suction.

Column packing The column should be homogeneously packed with care, following the instructions of the manufacturer. A Superformance column (a length of 5 cm and an inner diameter of 1 cm; Merck, Darmstadt, Germany) is well suited for this type of chromatography.

1. Before packing the Superformance 50–10 column, rinse the glass tube with water.

2. Mount the column, including the lower adapter and the filling tube, according to the instruction manual. Insert the lower filter element (filter F 20 μm), using the filter insertion tool, before mounting the filling tube.

3. Prepare a homogeneous gel slurry by mixing 6 ml of sucked-off gel with 6 ml of 1 M NaCl solution.

4. Pour the slurry into the vertically mounted column with the column outlet closed. Use a glass rod to pour water into the filling tube in order to fill the residual space above the gel slurry.

5. Connect the filling tube immediately to the pump, open the column outlet and start the pump at a flow rate of 5–7 ml/min, delivering equilibration buffer.

6. Allow the gel to pack completely within 10–15 min, then stop the pump and close the column outlet.

7. Decant or suck off the supernatant from the gel bed and disconnect the filling tube.

8. Rinse the column end piece with water; remove excess gel so that the gel surface has a distance of 5–10 mm to the edge of the glass tube.

9. Use the insertion tool to insert the upper polypropylene filter plate (filter F, 20 µm) on top of the gel and rinse the space above the filter with water. No gel particles should remain above the filter plate.

10. Mount the upper adapter, thereby compressing the gel bed.

11. Equilibrate the packed gel with at least 3–5 bed volumes of the equilibration buffer.

The performance of the packed column can be checked by running test chromatograms. The basic characteristics, i.e., peak symmetry, number of theoretical plates per meter of bed height (N) of the column, can be obtained injecting 0.1 ml of 0.3% acetone with a flow rate of 1 ml/min and monitoring the eluate at 280 nm. If the gel bed shrinks, loosen the counter screw and retighten the adapter.

A mixture of peptides (neurotensin, bradykinin, angiotensin, somatostatin and bombesin) is taken as an example to demonstrate efficient separation of the components of the mixture (see Fig. 1). Fractogel EMD 650 (S) chelate was packed into a Superformance column with a dimension of 50 x 10 mm with a bed height of about 5 cm as described above. For purification of other proteins and/or peptides the same conditions can be tried:

Performance of IMAC

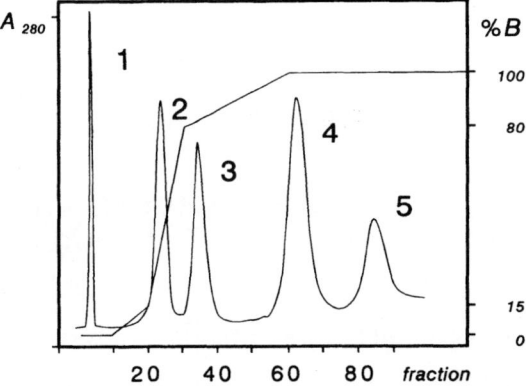

Fig. 1. Chromatography of peptides on an IDA-Cu(II) column. Neurotensin, bradykinin, angiotensin III, somatostatin and bombesin were applied on an IDA-Fractogel EMD chelate column charged with Cu(II). The column was equilibrated with 0.02 M sodium phosphate, 1 M NaCl (pH 7). After washing, the column was developed with a pH gradient as indicated using 0.1 M sodium phosphate, 1 M NaCl (pH 3.8) with a flow rate of 1 ml/min. Neurotensin (*peak 1*) does not bind to the column and elutes during the washing step. The strongest binding was found for the histidine containing peptides bombesin (*peak 4*) and angiotensin III (*peak 5*). The serine containing peptides bradykinin (*peak 2*) and somatostatin (*peak 3*) display a weaker affinity

1. Equilibrate the packed column with buffer A: 20 mM sodium phosphate buffer, 1 M NaCl (pH 7). Perform all of the following steps at a flow rate of 1 ml/min.

2. Subsequently wash with 2 column volumes of 1 M NaCl solution to remove the phosphate completely.

3. Load the column with 250 mM $CuSO_4$ diluted in 0.1 M sodium acetate buffer (pH 6) by pumping the solution through the column. One bed volume of metal salt solution is sufficient.

4. Wash with 2 column volumes 1 M NaCl solution to remove unbound metal ions.

5. Reequilibrate with 2 column volumes of buffer A.

6. Prepare the peptide samples with the following concentrations:

    – Dissolve 0.1 mg neurotensin in 1 ml of buffer A.

    – Dissolve 0.1 mg bradykinin in 1 ml of buffer A.

    – Dissolve 0.2 mg angiotensin III in 1 ml of buffer A.

    – Dissolve 0.2 mg somatostatin in 1 ml of buffer A.

    – Dissolve 0.4 mg bombesin in 1 ml of buffer A.

7. Mix the final sample solution by pipetting 0.5 ml of each of the above solutions together.

8. Apply 2.5 ml of the final sample mixture onto the column and wash with buffer A until the base line is constant. Neurotensin will not bind to the column and elutes in the first peak after 10 min (Fig. 1).

9. Elute the remaining peptides using buffer B: 100 mM sodium phosphate buffer, 1 M NaCl (pH 3.8). The gradient should be:

    – 0%–15% buffer B in 10 min

    – 15%–80% buffer B in 20 min for the elution of bradykinin,

    – 80%–100% buffer B in 30 min for the elution of somatostatin

    – 100% buffer B in 40 min for the elution of bombesin and angiotensin

Alternative type of elution For example, a 0.1 M sodium acetate buffer containing 0.5–1 M NaCl (pH 3.0) can also be used. Alternatively, bound proteins can be eluted from the column using stepwise descending pH gradients generated with salt-containing phosphate buffers adjusted to pH 6.0 or pH 4.0. It is also possible to develop an IMAC column with an imidazole gradient. For this, the column should be equilibrated prior to use with buffer to which 1 mM imidazole has been added, e.g., 20 mM sodium phosphate buffer, 0.5 or 1 M NaCl, 1 mM imidazole (pH 7.5). The actual elution then occurs with a gradient of 1–200 mM imidazole in the equilibration buffer. The flow rate can be increased to about 5 ml/min for columns with an inner diameter of 1 cm (linear flow rate about 380 cm/h).

For the regeneration of the metal chelate columns, approximately 1 bed volume of a 0.1–1 M HCl solution at a flow rate of 0.5 ml/min has to be pumped through the column. This removes all metal ions. After rinsing with 2 column volumes of 50 mM phosphate buffer, 1 M NaCl (pH 7), the column can be loaded with metal ions again. Fractogel EMD chelate is also stable against alkali treatment, so that the column can be cleaned with 0.5 M NaOH. For storage, a 20% ethanol solution prepared with equilibration buffer should be used.

Regeneration, storage

- Since a high salt concentration has to be used, precipitation of proteins may occur. Before chromatography is started, precipitated compounds must be removed by filtering the sample or by spinning down the samples in a benchtop centrifuge.

- If binding of the molecule of interest is too strong, another metal salt may give better results. The loading step should thus be performed using a 250 mM solution of either $NiCl_2$, $CoCl_2$ or $ZnCl_2$; $MnCl_2$ or $MgCl_2$ can also be useful.

- Sometimes it is necessary to change the buffers slightly and to add substances, thereby stabilizing the protein of interest. As an example, for the purification of glucokinase from yeast, a buffer A with the following composition was successfully used: 20 mM phosphate buffer, 1 M KCl, 10 mM glucose (pH 7.5). IMAC was performed using immobilized cobalt ions. The enzyme was eluted within a single peak by a decreasing imidazole gradient generated with buffer A, which was adjusted to pH 4.0 (Jacob et al. 1991).

Trouble-shooting

The following protocol can be used as a standard procedure for the isolation of antibodies by TAC (see Fig. 2). Fractogel EMD TA (S) is packed into a Superformance column with a column dimension of 50 x 10 mm with a bed height of about 5 cm according to the procedure described above.

Performance of TAC

A 280

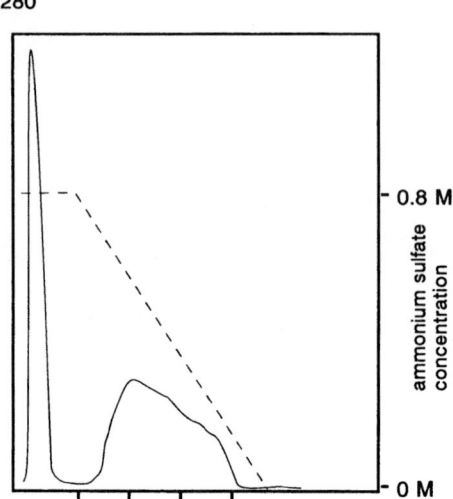

Fig. 2. Thiophilic adsorption chromatography of antibodies on Fractogel EMD TA 650 (S). A sample containing albumin and immunoglobulins was applied to a Fractogel TA column with dimensions of 50 x 10 mm previously equilibrated with 20 mM sodium phosphate buffer containing 0.8 M ammonium sulfate (pH 7). Elution was achieved by a decreasing ammonium sulfate gradient generated with buffer B, consisting of 20 mM sodium phosphate without any ammonium sulfate, at a flow rate of 1 ml/min. The first peak represents the nonbound albumin fraction; the polyclonal antibodies can be eluted during the falling salt gradient

1. Equilibrate the column with 5 bed volumes of buffer A: 20 mM phosphate buffer containing 0.8 M $(NH_4)_2 SO_4$ (pH 7). The flow rate should be 2 ml/min. Perform all of the following steps at a flow rate of 1 ml/min.

2. Prepare solutions of human γ-globulin and human serum albumin with the following concentrations:

   - Dissolve 2 mg of human γ-globulin (Serva, Heidelberg, Germany) in 1 ml of buffer A.

   - Dissolve 5 mg of human serum albumin (Sigma, Heidelberg, Germany) in 1 ml of buffer A.

3. Mix the final sample solution by pipetting 0.1 ml of each of the above solutions together.

4. Inject 0.2 ml of the final sample mixture onto the column

5. Subsequently wash with buffer A to remove unbound protein (albumin) until the UV baseline is constant.

6. Elute the bound proteins with a 30 ml linear gradient of 0.8 M–0 M ammonium sulfate in buffer A (i.e., 0%–100% B in 30 min at a flow rate of 1 ml/min, where buffer B is 20 mM phosphate adjusted to pH 7 containing no ammonium sulfate).

7. After elution of antibodies wash with 2 column volumes of buffer B.

8. Regenerate the column with 2 column volumes of a 0.1 M NaOH solution at a flow rate of 0.5 ml/min.

Alternatively, bound proteins can be eluted from the column using a linear gradient of decreasing ammonium sulfate concentration in combination with an increasing linear gradient of ethylene glycol concentration (0%–50%).The flow rate can be increased to about 5 ml/min for columns with an inner diameter of 1 cm (linear flow rate about 380 cm/h), although the capacity will slightly decrease. *(Alternative type of elution)*

Short-term treatment with NaOH is best suited for the regeneration of Fractogel EMD TA. Rinse the column with 2 bed volumes of 0.1 M NaOH solution at a flow rate of 0.5 ml/min. Regeneration can also be performed by rinsing with 50% ethylene glycol. Another successful method to remove tightly bound or denatured material from the column is to rinse with 6 M urea. For regeneration with organic solvents, a linear flow rate of 1 cm/min should not be exceeded. A 20% ethanol solution (v/v) is recommended for storage of the column. *(Regeneration, storage)*

References

Azzi A, Bill K, Broger C (1982) Affinity chromatographic purification of cytochrome c binding enzymes. Proc Natl Acad Sci USA 79:2447–2450

Belew M, Juntti N, Larsson A and Porath J (1987) A one-step purification method for monoclonal antibodies based on salt-promoted adsorption chromatography on a thiophilic adsorbent. J Immunol Methods 102:173–182

Bischoff FR, Klebe C, Kretschmer J, Wittinghofer A, Ponstingl H (1994) RanGAP1 induces GTPase activity of nuclear Ras-related RAN. Proc Natl Acad Sci USA 91:2587–2591

Carlsson J, Janson JC, Sparrman M (1989) Affinity chromatography. In:Janson JC, Ryden L (eds) Protein purification. VHC, New York, pp 275–329

Dean PDG, Johnson WS, Middle FA (1985) Affinity chromatography. IRL, Oxford

Eriksson KO (1989) Hydrophobic interaction chromatography. In:Janson JC, Ryden L (eds) Protein purification. VHC, New York, pp 207–226

Hansen P, and Lindberg GJ (1992) Immobilized metal ion affinity chromatography of synthetic peptides - binding via the alpha-amino group. J Chromatogr 627:125-135

Hansen P, Lindberg GJ (1995) Importance of the alpha-amino group in the selective purification of synthetic histidine peptides by immobilized metal ion affinity chromatography. J Chromatogr A 690:155-159

Hemdan ES, Zhao YJ, Sulkowski E, Porath J (1989) Surface topography of histidine residues: A facile probe by immobilized metal ion affinity chromatography. Proc Natl Acad Sci USA 86:1811-1815

Hermanson GT, Mallia AK, Smith PK (1992) Immobilized affinity ligand techniques. Academic, London

Hutchens TW, Porath J (1986) Thiophilic adsorption of the immunoglobulins-Analysis of conditions optimal for selective immobilization and purification. Anal Biochem 159: 217-226

Jacob L, Beecken V, Rose M, Entian KD, Bartunik LJ, Bartunik HD (1991) Purification of yeast glucokinase by immobilized metal affinity chromatography. Biological Chemistry Hoppe-Seyler 372:683

Kagedal L (1989) Immobilized metal ion affinity chromatography. In:Janson JC, Ryden L(eds) Protein purification. VHC, New York, pp 227-251

Mohr P, Pommerening K(1985) Affinity chromatography. Marcel Dekker, New York

Narayan SR, Crane LJ (1990) Affinity chromatography supports: a look at performance requirements. Trends in Biotechnology 8:12-16

Peters JH, Baumgarten (eds.) (1989) Monoklonale Antikörper, Springer, Berlin, Heidelberg, New York, London

Porath J, Carlsson J, Olsson I, Belfrage G (1975) Metal chelate affinity chromatography, a new approach to protein fractionation. Nature 258:598-599

Porath J, Maisano F, Belew M (1985) Thiophilic adsorption – A new method for protein fractionation. FEBS Lett 185:306-310

Porath J, Olin B (1983) Immobilized metal ion affinity adsorption and immobilized metal ion affinity chromatography of biomaterials. Serum protein affinities for gel-immobilized iron and nickel ions. Biochemistry 22: 1621-1630

Regnier FE (1991) Perfusion chromatography. Nature 350:634-635

Schulze R, Kontermann R, Queitsch I, Dübel S, Bautz E (1994) Thiophilic adsorption chromatography of recombinant single-chain antibody fragments. Analytical Biochemistry 220:212-214

Scopes RK (1987) Dye-ligand and multifunctional adsorbents: an empirical approach to affinity chromatography. Analytical Biochemistry 165: 235-246

Scouten WH (1981) Affinity chromatography. Wiley, London

Sulkowski E (1985) Purification of proteins by IMAC. Trends in Biotechnology 3:1-7

Tijssen P (1985) Practice and theory of enzyme immunoassays. Elsevier, Amsterdam, pp 43-78

Wilcheck M, Bayer EA (1990) Avidin-biotin technology. In: Methods Enzymol, vol. 184 , Academic, London, pp 5-45

Wilchek M, Miron T, Kohn J. (1984) Affinity chromatography. In: W.B. Jakoby (ed.) Methods Enzymol, vol 104 Academic, London, pp 3-55

Wong SS (1991) Chemistry of protein conjugation and cross-linking. CRC, Boston

Chapter 3

Protein Purification by Aqueous Two-Phase Systems

H. SCHÜTTE

3.1
Introduction

Liquid-liquid partition is a highly sophisticated and well-established separation technique in organic chemistry. However, organic solvents cannot be used for the purification of microbial biologically active proteins, as they will either denature the proteins or will not dissolve them. Instead, partition of biologically active proteins can be performed in aqueous two-phase systems (described in detail by Albertsson 1986), which consist of up to 90% water in both phases. The phases are formed as a result of the incompatibility of certain water miscible polymers. If a solution of a highly water miscible polymer, e.g., polyethylene glycol, is mixed with a solution of a second highly water miscible polymer, e.g., dextran, one may obtain two-phases depending on the concentrations of the polymers. Aqueous phase systems can also be achieved using one polymer and an appropriate salt (e.g., potassium phosphate). In Fig. 1 the phase diagrams of polyethylene glycol/potassium phosphate and polyethylene glycol/dextran systems (Albertsson 1986; Kroner et al. 1982) are presented. It can be seen that below the binodal curve, a homogeneous solution will be obtained; but, if certain concentrations are exceeded, phase separation takes place and a PEG-rich upper phase and a salt-rich lower phase are formed. These are no longer miscible with each other, despite the fact that both phases contain a high proportion (> 75%) of water. All mixtures with compositions represented by points of the tie-line from T through M to B (Fig. 1a) will yield phases with identical compositions of upper and lower phase but different phase volumes. Systems of the same tie-line therefore exhibit identical partition coefficients. At the critical point C, where the addition of a minute amount of water transforms a two-phase system to a homogeneous solution, the two phases should theoretically have identical compositions

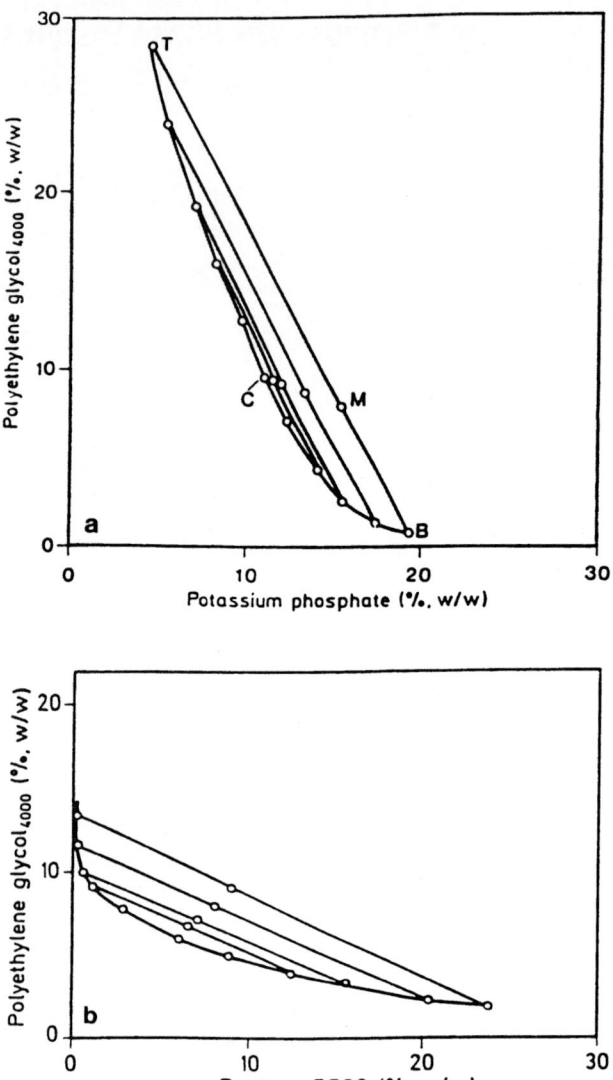

Fig. 1. a Phase diagram of the system polyethylene glycol 4000/potassium phosphate at 20 °C. *C*, critical point; *T*, composition of the upper phase; *B*, composition of the lower phase; *M*, composition of the total phase. **b** Phase diagram of the system polyethylene glycol 4000/dextran T-500 at 20 °C. (From Albertsson 1986)

and volumes and give partition coefficients of 1. The partition coefficient is defined by Eq. (1):

$$K = \frac{C_T}{C_B} \tag{1}$$

where C_T and C_B are the equilibrium concentrations of the partitioned compounds in the top and bottom phases, respectively. The partition coefficient of biologically active proteins and enzymes is constant for a given system over a fairly wide range of concentrations. Any molecule will accumulate in the phase in which a maximum number of interactions are possible and partition in such way that a minimum of the energy content of the system is reached. The Brönstedt equation:

$$K = e^{\frac{\lambda \cdot M}{k \cdot T}} \tag{2}$$

where λ is a parameter characterizing the phase system and interactions with the compound of interest, M is the molecular weight, k the Boltzmann constant and T the absolute temperature, describes quantitatively – λ values are unknown – an exponential dependence of the partition coefficient on the molecular weight and the factor λ. If dealing with compounds possessing very high molecular weights, e.g., $M > 10^6$ dalton, a one-sided partition can produce drastic changes of the partition coefficient.

Salts have a pronounced effect on the partition of proteins, which is mainly determined by the kind and ratio of ions present and by their different affinities for the phases. This uneven distribution of salt ions is the reason for the electrostatic potential difference between the phases. For a salt, the ions of which have the charges Z^+ and Z^-, the electrostatic potential is given by (Albertsson et al. 1990):

$$\psi = \frac{R \cdot T}{(Z^+ + Z^-) \cdot F} \ln \frac{K_-}{K_+} \tag{3}$$

where R is the gas constant, T is the absolute temperature, Z^+ and Z^- are the charges of the ions, F is the Faraday constant, K^+ and K^- are the hypothetical partition coefficients of the ions in the absence of a potential and Ψ is the interfacial potential. The larger the ratio of K^+ to K^-, the larger will be the interfacial potential. In the presence of excess of salt a protein will partition according to

Table 1. Parameters influencing partition in aqueous two-phase systems

Types of polymers composing the two-phase systems
Average molecular weight of the polymers
Molecular weight distribution of polymers
Length of the tie-line (complex function of concentration)
Types of ions composing or added to the system
Ionic strength
pH
Temerature

$$\ln K_p = K_p^{\circ} + \frac{F \cdot Z}{R \cdot T} \cdot \psi \qquad (4)$$

where K_p is the partition coefficient of the protein, K_p° the value of K_p when the interfacial potential (generated by the excess salt) is zero or when the protein net charge is zero, and Z is the net charge of the protein.

The difference in partition of the ions of the salt generates an interfacial potential according to Eq. (3), which in turn affects the partition of the protein according to Eq. (4).

The partition of proteins and other compounds in aqueous two-phase systems is influenced by a large number of parameters. The important variables are summarized in Table 1. Most, if not all, of the parameters listed do not act independently, therefore calculation or theoretical predictions of the partition coefficient for a given protein cannot be carried out at present. Suitable conditions for a desired partition have to be found by experimentation. This is aided considerably if fast analytical determination of the compound of interest is available. The experimental reproducibility of partition coefficients is normally in the range of around 5% for any volume analyzed. The basic aspects of the method as well as different applications have been discussed in two books (Albertsson 1986; Walter et al. 1985) and several review articles (Kula et al. 1982; Hustedt et al. 1985 a,b).

3.2
Removal of Cell Debris from Microorganism Homogenates

To achieve separation between intracellular proteins and cell debris after cell disruption, conditions have to be established so that the protein of interest and cell debris prefer opposite phases. At the same time, however, the partition coefficient of the protein and the volume ratio of the phases should be sufficient to extract the protein in high yield. In PEG/salt systems it is rather simple to partition the cell debris into the salt-rich lower phase. For example, in a system made up of 18% (w/w) PEG 1540, 7% (w/w) potassium phosphate, and 20% (w/w) broken wet cells, the cell debris of many organisms, yeast as well as bacteria, partitions quantitatively into the lower phase. The 18/7 mixture is clearly below the binodal curve of the carrier system (see Fig. 1), indicating that polymers from the broken cells contribute to the phase formation, as expected.

Partitioning of the cell debris into the lower phase does not necessarily mean that the desired protein partitions to the upper phase. This is often much more difficult to achieve. No general rules can be given for this problem, which must be solved empirically. The important parameters are the molecular weight of the PEG, the pH of the system, the concentration of salt and cell homogenate, and other factors as listed in Table 1. By adequate variation and combination of these parameters, suitable conditions for extraction processes in all cases investigated so far have been found. Typical data for the extraction of a number of enzymes from disrupted cells of various microorganisms are summarized in Table 2, which may be summarized as follows:

- In most cases practical yields of more than 90% could be obtained in a single extraction step.

- The partition coefficients are mainly in the range of 2–20.

- In many cases a considerable amount of contaminating protein, up to 90%, is separated, together with the cell debris, from the desired product.

- In most cases PEG/salt systems are used.

- The concentration of cell homogenate in the phase system is mainly in the range of 20%–30% (w/w).

The removal of contaminating protein together with the cell debris is an important aspect of the economics of such processes. In this way , one or two initial purification steps can be omitted; these steps are commonly employed after cell debris separation by filtration or centrifugation. In addition, most of the nucleic acids and polysaccharides may also be removed at the first extraction step. Another important aspect is the enhanced stabilization of enzyme activities by the polymer(s) used. Partition may be carried out at room temperature without any cooling.

Table 2. Extraction of enzymes from cell homogenates

Organism	Enzyme	Constituent of the phase system	K_{enzyme}	Yield (%)
Candida boidinii	Formaldehyde dehydrogenase	PEG 4000/crude dextran	10.8	94
	Formate dehydrogenase	PEG 4000/crude dextran	7.0	91
	Formate dehydrogenase	PEG 1000/potassium phophate	4.9	94
	Isopropanol dehydrogenase	PEG 1000/potassium phosphate	18.8	98
Saccharomyces carlsbergensis	α-Glucosidase	PEG 4000/dextran T-500	1.5	75
Saccharomyces cerevisiae	α-Glucosidase	PEG 4000/dextran T-500	2.5	86
	Glucose-6-phosphate dehydrogenase	PEG 1000/potassium phosphate	4.1	91
Streptomyces species	Glucose isomerase	PEG 1550/potassium phosphate	3.0	86
Klebsiella pneumoniae	Pullulanase	PEG 4000/dextran T-500	2.96	91
	Phosphorylase	PEG 1550/dextran T-500	1.4	85
Escherichia coli	Isoleucyl tRNA synthetase	PEG 6000/potassium phosphate	3.6	93
	Fumarase	PEG 1550/potassium phosphate	3.2	93
	Aspartase	PEG 1550/potassium phosphate	5.7	96
	Penicillin acylase	PEG 4000/crude dextran	1.7	90
Bacillus sphaericus	Leucine dehydrogenase	PEG 4000/crude dextran	9.5	98
Bacillus species	Glucose dehydrogenase	PEG 4000/crude dextran	3.2	95
Brevibacterium ammoniagenus	Fumarase	PEG 1550/potassium phosphate	0.24	90
Lactobacillus species	Lactate dehydrogenase	PEG 4000/dextran PL-500	6.3	95

3.3
How To Find Suitable Partition Conditions

In general, screening experiments are carried out on a 10 g scale. Stock solutions of the polymers to be used (Dextran 20%, PEG 50%, salt 40%) have to be prepared on a weight per weight basis. For example, variable amounts of stock solution of PEG 1540 and salt must be weighed and placed in 10 ml graduated centrifuge tubes. The PEG and salt can be added in solid form or, as described already, as a concentrated solution. The cell homogenate is added and the pH is controlled and fixed. The system is then brought to a final total weight of 10 g by the addition of water. Generally, partition studies must be carried out at room temperature. Mixing is performed on a vortex agitator (around 3 min) or by a tube rotator (20–30 min at 5 rpm). The latter mode of mixing is recommended when large amounts of the components have been put into the tube in solid form. Phase separation is performed by centrifugation at 2000 g for 5 min in a swinging bucket rotor. The phase volumes are noted and centrifugation is repeated to control for complete phase separation. If a real phase system has been formed the volume ratio remains constant after the first centrifugation. The volumes are noted and samples from the upper phase are removed with a pipet and analyzed for the concentration of the desired product. If of interest or if necessary, the concentration in the viscous lower phase is also analyzed. The partition coefficient K of the desired protein can be determined by dividing the observed activity (U/ml) in the upper phase by the activity (U/ml) in the lower phase.

As an example the whole procedure will be demonstrated for the extraction of fumarase from *S. cerevisiae*.

Cells of *S. cerevisiae* must be suspended in 100 mM potassium phosphate buffer, pH 7.5, representing a 40% (w/v) cell suspension. The pH of the suspension must be controlled and adjusted. Afterwards the cells can be disrupted by two consecutive passes through a continuously operating agitator bead mill (Schütte and Kula 1990a) or by a high-pressure homogenizer at 100 MPa (Schütte and Kula 1990b). After cell disruption the pH of the homogenate must be readjusted to 7.5. The enzyme activity (U/ml) and the protein content (mg/ml) must be measured using standard assay procedures and the specific activity (U/mg) of the homogenate calculated.

A standard 10 g system, which is commenly handled in the laboratory for experimental trials, contains (example for fumarase):

Margin notes:

Screening experiments

Extraction of fumarase from Saccharomyces cerevisiae

50%	(w/w)	of the homogenate (40% w/v)	=	5.0 g
17%	(w/w)	polyethyene glycol (PEG$_{1500}$)	=	1.7 g
7%	(w/w)	potassium phosphate salt (pH 8.0)	=	0.7 g
26%	(w/w)	deionized water	=	2.6 g
100%			=	10.0 g

For calculation of the total units (100% value) given to the aqueous phase system it is important to notice that 5 g of homogenate is not 5 ml; normally it is 4.7 or 4.8 ml.

After mixing and centrifugation as described above the following data from the laboratory trials have to be determined:

- Concentration of the enzyme of interest in both phases

- Determination of the protein content in both phases

- Volume of upper (top) and lower (bottom) phases

- pH value in both phases

- Total units and total mg in both phases

- yield, specific activity and purification factor of the desired enzyme in the upper phase

If the volume ratio of the phase system is not optimal (2/3 upper phase and 1/3 lower phase) or the yield of the desired enzyme in the upper phase is below 90% the following parameters can be varied for further optimization:

- Amount of homogenate

- Amount of PEG 1540

- Amount of potassium phosphate

- pH of the salt (mixture of K_2HPO4 and KH_2PO4)

In principle it is sufficient to analyze the upper phase only, because the salt-rich lower phase containing the cell debris, etc., is going to waste and the PEG-rich upper phase should contain the desired enzyme.

The enzyme containing phase cannot be used for common chromatographic techniques directly because its viscosity is too high. Therefore the PEG introduced into the enzyme solution to establish the phase system must be removed later on. A number of methods are available for this purpose. One simple way is to produce a new phase system by adding salt to the PEG-rich upper phase and shifting the desired enzyme to the salt phase, as discussed below.

3.4
Further Purification by Subsequent Partition Steps

The protein which was partitioned into the upper phase at the first extraction step and separated by centrifugation or by a separator can be further purified by the addition of suitable concentrations of salt to form a secondary two-phase system. Thus, the major part of the PEG and also a part of the salt (if the first system was a PEG/salt system) is reused. The protein of interest should be shifted at this stage into the lower salt-rich phase (e.g., by changing the pH or adding small amounts of sodium chloride) in order to separate it from the bulk of the viscous PEG. Separation of these two-phases can be carried out in a simple settling tank. Settling times between 30 and 90 min can be sufficient to obtain apparently clear phases. Separation under gravity can be carried out at any scale without difficulty and can be performed continuously in a suitable settling tank or in a mixer settler device when multistage procedures are desirable (Hustedt et al. 1985b).

To shift the enzyme fumarase from the PEG-rich upper phase into the salt-rich lower phase the following system can be used:

60 %	(w/w)	upper phase I	=	6 g
7 %	(w/w)	potassium phosphate salt mixture (pH 7.0)	=	0.7 g
0.5 %	(w/w)	sodium chloride	=	0.05 g
32.5 %	(w/w)	deionized water	=	3.25 g
100 %			=	10.0 g

The following parameters can be varied for further optimization:

- Amount of upper phase I
- Amount of potassium phosphate salt
- pH of the salt mixture
- Amount of sodium chloride
- Addition of PEG 10000

After mixing and separation of the phases as described above, both phases have to be analyzed. The K value should be at least in the range of 0.01 to yield a high amount of total fumarase in the lower salt-rich phase. The experimental data (10 g scale) for the extraction of fumarase from *S. cerevisiae* are summarized in Table 3. A general extractive purification scheme for a two-step

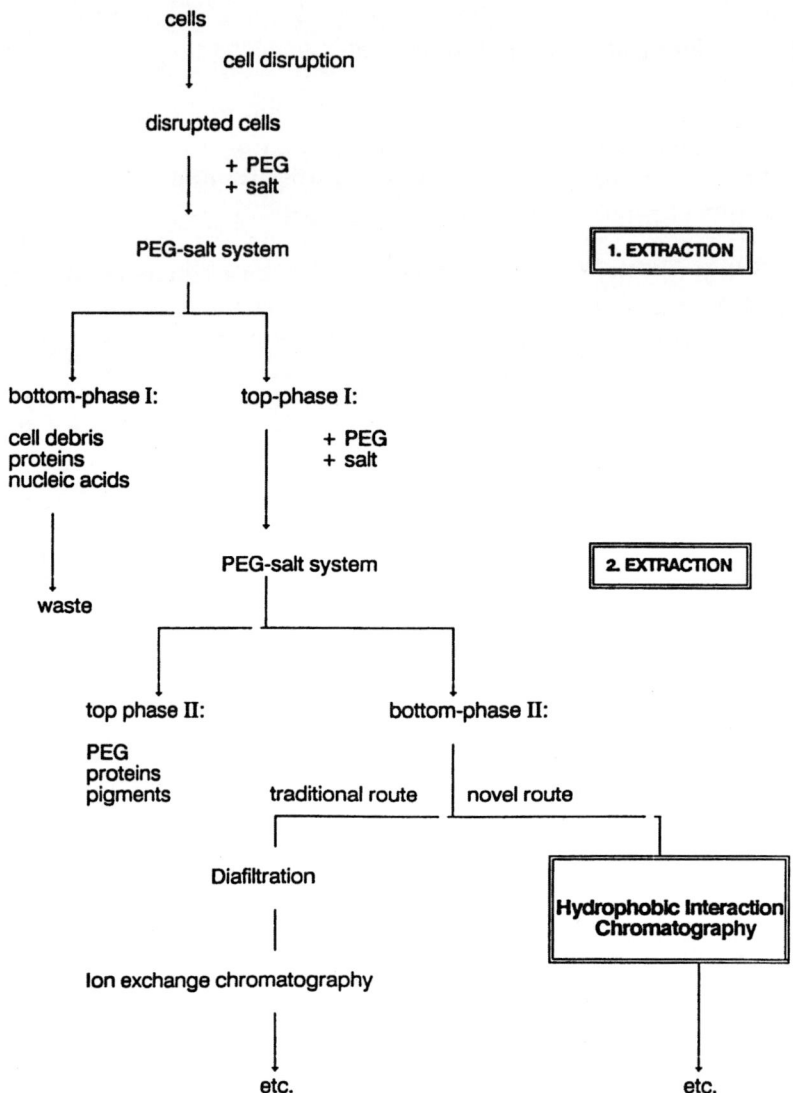

Fig. 2. General scheme of extractive protein purification followed by chromatographic techniques

liquid-liquid extraction process is shown in Fig. 2. Data showing the efficiency of the purification for some proteins are summarized in Table 4.

Table 3. Examples of enzymes partially purified by one or several subsequent partition steps

Enzyme	Source	Partition steps (N)	Overall purification factor	Overall yield (%)
Pullulanase	Klebsiella pneumoniae	4	6.3	70
α-1,4-glucan phosphorylase	Klebsiella pneumoniae	2	2.5	81
Fumarase	Brevibacterium ammoniagenes	3	8.8	70
Aspartase	Escherichia coli	3	23	68
Glucose dehydrogenase	Bacillus species	3	33	83
Leucine dehydrogenase	Bacillus sphaericus	2	7.7	97
Formate dehydrogenase	Candida boidinii	4	4.4	70
Glucose isomerase	Streptomyes species	1	2.5	86
β-glucoisidase	Lactobacillus cellobiosus	1	2.4	98
Interferon	Human fibroblasts	1	350	86

The first partition step always aimed to remove broken cells (with the exception of interferon).

Table 4. Primary purification of fumarase from *Saccharomyces cerevisiae* by liquid-liquid extraction

Step	Volume (m)	Activity (U/ml)	Total activity (U)	Protein (mg/ml)	Total protein (mg)	Specific activity (U/mg)	Yield (%)	Purification (fold)
Crude extract	4.8	14.6	70.1	39.2	188.2	0.37	100	1
1. Extraction upper phase	6.5	10.4	67.6	12.1	78.7	0.86	96.4	2.3
2. Extraction lower phase	4.9	12	58.8	8.9	43.6	1.35	83.9	3.6

3.5
Scale-Up of Aqueous Two-Phase Systems

The scale-up of extractive enzyme recovery with high precision is easy and straightforward. This is due to the fast attainment of phase equilibrium, the simplicity of the technical operations involved, and the commercial availability of the necessary equipment. Extractive enzyme recovery has been carried out on scales varying from 10 g to more than 1 ton of phase system. For scaling up from 10 g the amounts of all components can be proportionally increased, leading to identical results, within experimental error, for the extraction process. This is shown, for example in the purification of leucine dehydrogenase from Bacillus cereus (Schütte et al. 1985)

A total of 30 kg *B. cereus* cells were disintegrated using a continuously operating industrial agitator mill, followed by heat treatment for 10 min at 63 °C. For extraction of *B. cereus* homogenate, solid PEG 1540 and solid potassium phosphate salt (pH 8.0 mixture) were added and stirred for 1 h to form a PEG/salt system and ensure equilibrium of partition. The whole extraction system consisted of 18% (w/w) PEG 1540 (25.2 kg), 7% (w/w) potassium phosphate salt (9.8 kg), 70 l cell homogenate, and deionized water to give a 140 kg (128 l) phase system. Liquid-liquid separation was carried out employing a disc stack separator (Westfalia separator model SAOH-205, Westfalia AG, Oelde, Germany) with a feed rate of 100 L/h. Some 98% of the L-leucine dehydrogenase was found in the PEG-rich upper phase I (86 l), which is essentially free from any solid material. After this step the specific activity of the enzyme was increased 11 times compared to the activity in the cell homogenate. This is about 2.2 times higher than the activity reported for the conventional centrifugation step.

Purification of leucine dehydrogenase from Bacillus cereus

The second phase system had a final volume of 172 l and contained 86 l of the upper phase I, 11% (w/v) potassium phosphate salt (pH 7.0 mixture) and sodium chloride to a final concentration of 0.3 M. After stirring for 1 h, separation of the phases was allowed to proceed under gravity in a settling tank. Complete phase separation could be accomplished within 30 min. After separation the lower phase II was harvested and diafiltrated against 10 mM potassium phosphate buffer, pH 7.5. Thereafter, further purification of the enzyme was started as summarized in Table 5 (Schütte et al. 1985).

Table 5. Purification of L-leucine dehydrogenase

Step	Volume (l)	Protein (g)	Total activity (U · 10⁻³)	Specific activity (U/mg)	Yield (%)	Purification (fold)
Crude extract	70	723	434	0.6	100	1
Heat treatment	70	145	421	2.9	97	4.8
Upper phase I	86	61	412	6.8	95	11.3
Lower phase II	99	40	373	9.3	86	15.5
Diafiltration	2.4	22	360	16	83	26.7
DEAE-Sephacell	4.7	13	308	24	71	40
Sephacryl S-200	0.48	6.3	260	41	60	68.3
Sepharose 4B	1.4	3.6	208	57	48	95

Oxidative deamination assay.

Hart et al. (1994) have published the procedure for large scale, in situ isolation of periplasmic human insulin-like growth factor I (IGF-I) from *E. coli*. IGF-I accumulates in both folded and aggregated forms in the fermentation medium and cellular periplasmic space when expressed in *E.coli* with an endogenous secretory signal sequence. Due to its heterogeneity in form and location, a low yield of IGF-I was obtained using a typical refractile body recovery strategy. To enhance recovery yield, IGF-I was solubilized and extracted from cells while still in fermentation broth. This method, called in situ solubilization, involves addition of chaotrope (final concentration 2 M urea) and reductant (final concentration 10 mM dithiothreitol) to alkaline fermentation broth and provides recovery of about 90% of all IGF-I in an isolated supernatant. An aqueous two-phase extraction procedure was developed which partitions soluble non-native IGF-I and biomass solids into separate liquid phases. This two-phase extraction procedure involves addition of polymer (PEG 8000) and salt (sodium sulfate) to the solubilization mixture and provides about 90% recovery of solubilized IGF-I in the upper phase. The performance of the solubilization and aqueous extraction procedures is reproducible at scales ranging from 10 to 1000 l and provides a 70% cumulative recovery yield of IGF-I in the isolated upper phase. Together, the techniques of in situ solubilization and aqueous two-phase extraction provide a new, high yield approach for isolating recombinant protein which has accumulated in more than one form during fermentation.

3.6
Continuous Cross-Current Extraction

The extractive purification procedures summarized in Table 3 have been performed batchwise with several, subsequent, single-stage extraction steps. However, Hustedt et al. (1985b, 1987) have designed a plant for continuous cross-current extraction. The processing of aqueous-phase systems in large-scale protein recovery necessitates two technical operations: (1) mixing of the phase components followed by phase dispersal; and (2) phase separation. Both can be performed continuously without difficulty. The general flow-scheme of enzyme purification by continuous cross-current extraction is shown in Fig. 3. Stock solutions of PEG and salts are injected into the process stream of cell homogenate. The mixture is passed through the first static mixer and from there through the first separator. The outflowing lower phase containing the cell debris and a major part of the contaminating protein goes to waste. The product containing upper

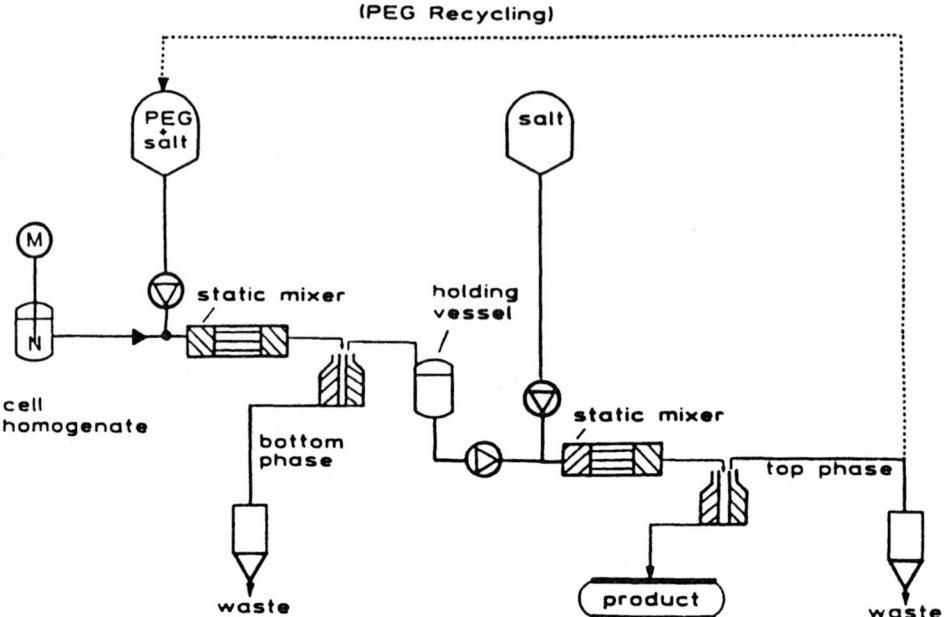

Fig. 3. Continuous cross-current extraction-flow scheme for protein purification. (From Hustedt et al. 1985b)

phase I is collected in a holding vessel when a separator with free outflow is used. From the holding vessel the upper phase I is pumped through the second static mixer before the mixer salt solution is added to the process stream in order to produce the secondary phase system. This is separated by the second separator. The outflowing lower phase (the salt-rich phase) now contains the product, while the PEG-rich upper phase may be recycled or else it goes to waste. Such a plant is capable of handling up to 200 kg biomass per day (8 h) in continuous operation controlled by a personal computer (Westfalia AG, Oelde, FRG), (Hustedt et al. 1987). Kroner et al. (1984) have discussed the economics of large-scale enzyme recovery and enrichment by liquid extraction in aqueous phase in comparison with other methods.

3.7
Conclusions

Aqueous two-phase systems are useful for separating a wide range of water-compatible substances (from peptides to cells). The selectivity of the separation normally increases with the size of the partitioned molecules or particles. The partition and separation capacity can be influenced by a number of parameters. Aqueous two-phase systems are well suited for large-scale purification of biomaterials such as enzymes and other biologically active proteins. This emerging technique is characterized by:

- High capacity

- Easy processing on any scale

- Easy and precise scale-up

- High product yields

- Low investment costs

- High potential for continuous processing

- Relatively large need for chemicals

In addition, no adverse effects regarding extraction have been observed during subsequent chromatographic separation (Fig. 2).

References

Albertsson PA (1986)Partition of cell particles and macromolecules; 2nd ed. Wiley, New York

Albertsson PA, Johansson G, Tjerneld F (1990) Aqueous two-phase separations. In: J.A. Asenjo (ed); Bioprocess technology Vol. 9. Separation processes in biotechnology. Marcel Dekker, New York, pp 287–327

Hart RA, Lester PM, Reifsnyder DH, Ogez JR, Builder SE (1994) Large scale, in situ isolation of periplasmic IGF-I from E.coli. Bio/Technology 12:1113–1117

Hustedt H, Kroner KH, Menge U, Kula MR (1985a) Protein recovery using two-phase systems. Trends Biotechnol. 3:139–144

Hustedt H, Kroner KH, Kula MR (1985b) Applications of phase partitioning in biotechnology. In: Walter H, Brooks DE, Fischer D (eds) Partitioning in aqueous two-phase systems: theory, methods, uses, and applications to biotechnology. Academic, Orlando, pp 529–582

Hustedt H, Börner B, Kroner, KH, Papamichael N (1987) Fully automated continuous crosscurrent extraction of enzymes in a two stage plant. Biotechnology Techniques 1:49–54

Kroner KH, Hustedt H, Kula MR (1982) Evaluation of crude dextran as phase-forming polymer for the extraction of enzymes in aqueous two-phase systems in large scale. Biotech Bioeng 24:1015–1045

Kroner KH, HustedtH, Kula MR (1984) Extractive enzyme recovery: economic considerations. Process Biochemistry 19:170–179

Kula MR, Kroner KH, Hustedt H (1982) Purification of enzymes by liquid-liquid extraction. In: A.Fiechter (ed) Advances in biochemical engineering, Vol. 24, Springer, Berlin, Heidelberg, New York, pp 73–118

Schütte H, Hummel W, Tsai H, Kula MR (1985) L-leucine dehydrogenase from Bacillus cereus: Production, large-scale purification and protein characterization. Microbiol.Biotechnol. 22: 306–317

Schütte,H, Kula MR (1990a) Bead mill disruption. In: JA Asenjo (ed) Bioprocess technology, Vol.9. Separation processes in biotechnology Marcel Dekker, New York, pp 107–141

Schütte H, Kula MR (1990b) Review: pilot- and process-scale techniques for cell disruption. Biotechnol. Appl. Biochem12: 599–620

Walter H, Brooks DE, Fisher D (1985) Partitioning in aqueous two-phase systems: Theory, methods, uses and applications to biotechnology. Academic Press, Orlando,

Rapid HPLC Separation of Proteins, Peptides, and Amino Acids Using Short Columns Filled with Nonporous Silica-Based Particles

F. GODT and R. M. KAMP

4.1
Introduction

For many years, silica-based chromatography supports have played an important role in HPLC due to the many features of silica. Advances in this technique have led to the development of nonporous silica-based particles 1.5 µm in diameter. The absence of pores eliminates intraparticle diffusion. The free available surface allows unhindered protein-matrix interaction and mass transfer, thus significantly improving sensitivity. In addition the short column, which has a highly polished inner surface, a stainless steel mesh and a fiber filter sandwich, along with the highly quality column packing technique results in shorter analysis times and therefore reduced solvent consumption. Faster separation minimizes the contact time of the biomolecules with both the solvent and the matrix, which often causes a loss of bioactivity or damage to the tertiary structure.

Small particles improve the efficiency of HPLC separation. The 30–33 mm columns packed with 1.5 µm silica particles provide the same theoretical number of plates (efficiency of a chromatographic column) as a 25 cm column packed with 5 µm material. The separation time can be reduced from hours to minutes. The sharp peaks improve detection limits and even traces of impurities are detectable.

The classical separation of biomolecules using porous silica causes nonspecific adsorption. The trapped molecules are eluted in later runs. Nonporous silica has no pores and nothing can be trapped on the surface; consequently, recoveries of up to 100% are possible. The nonporous columns have been developed for rapid separation of large biomolecules such as proteins (Unger 1986; Janzen 1987). When the temperature of the column is increased up to 80 °C, the separation can be shortened to even seconds (Kalghatgi 1987).

Nonporous supports have also been applied to peptide mapping (Kalghatgi 1987, 1989; Jilge 1987). As an example separation of cytochrome c tryptic peptides will be discussed below.

Application of nonporous columns to the separation of small molecules such as amino acids was investigated for the first time in 1993 (Stromer 1993). Fourteen amino acids were separated in 7 min using aqueous buffer with the addition of an ion pair-containing reagent and methanol as an organic modifier. For sensitive detection at the low picomole level, the amino acids were derivatized with o-phthaldialdehyde before being injected.

4.2
Eluents

4.2.1
Aqueous Buffer

Different aqueous buffers can be applied to the separation of proteins, peptides and amino acids, e.g., 0.1% trifluoracetic acid (TFA), ammonium acetate, ammonium formate or sodium phosphate. For protein and peptide separation volatile buffers at low salt concentration are preferable as they allow direct further investigation. Usually 0.1% TFA results in complete sample ionization and therefore sufficient resolution on reversed phase columns.

Preparation of aqueous eluents

1. Use deionized and ultrafiltrated water.

2. Prepare mobile phases in clean glassware.

3. For TFA buffers use sequence grade or three times distilled TFA.

4. For salt containing buffers add 1 mg sodium azide per 1 l buffer to prevent growth of microorganisms.

5. Filter the buffer using 0.45 μm filter.

6. Degas the aqueous buffer by sonication or, better, use a continuous degassing system, e.g., from ERC (Alteglofsheim, Germany).

7. Prepare fresh aqueous buffers every week.

4.2.2
Organic Modifiers

For the separation of proteins, peptides and amino acids by reversed phase systems different solvents can be used. Selection of organic modifiers guarantees low absorption at the wavelength used. For separation of large and hydrophobic proteins propanol is a stronger eluent and is superior to acetonitrile or methanol. Smaller molecules such as peptides show satisfactory separation using acetonitrile. Amino acids are usually separated by a gradient, consisting of aqueous buffer to methanol, which has characteristically less eluotropic strength than propanol and acetonitrile.

1. Use HPLC grade organic solvents only.

2. Use clean and dry glassware only.

3. Add 10% aqueous buffer to avoid trouble with an ascending baseline.

4. Degas the organic solvent by sonication or, better, use a continuous degassing system, e.g., ERC (Alteglofsheim, Germany).

Preparation of organic solvents

4.2.3
Sample Preparation

Proteins and peptides should be dissolved in the starting buffer in a concentration of 1–100 pmol/2–4 µl. The solution should not be contaminated with unsoluble particles. Diluted samples should be concentrated using a lyophilization system, with a centrifuge if possible, and should not be completely dried. To prevent problems in redissolving dried probes, it is possible to inject probes directly after enzymatic digestion or after a previous isolation step. For sufficient resolution as small a volume as possible should be injected. The sample loop of the injection valve should have a loop volume of 2–4 µl. A methanol washed, fine injection syringe should be used and the sample injected without air bubbles. For automatic injections, use microsample tubes to make sure that the solution is on the bottom of the tube.

4.3
Separation of Proteins and Peptides

4.3.1
Materials

HPLC system
- two HPLC pumps (model 510; Waters Co., Milford, MA)
- injection valve with a 4 μl sample loop (Rheodyne, Cotati, CA)
- gradient mixer with a mixing volume of 300 μl (BioRad, Hercules, CA)
- self-constructed column oven
- photodiode array detector (model 996; Waters Co., Milford, MA) – software Millenium 2.0 (Waters Co., Milford, MA)
- NPS 1.5 μm column (size: 33 x 4 mm) filled with 1.5 μm nonporous silica-based particles (Micra Scientific, Northbrook, IL, USA)
- Kovasil MS-H 1.5 μm column (size: 30 x 4 mm; a gift from Dr. Herbert Funke, Läutenring 29, 85235 Pfaffenhofen, Germany)
- inlet filter containing one 7 μm and one 3 μm stainless steel filter (Knauer GmbH, Berlin, Germany)

Reagents
- buffer A: 0.1% TFA in water: Filtered over a 0.45 μm membrane filter (Schleicher and Schuell, Dassel, Germany) and degassed continuously by a degasser (ERC model 3112; ERC, Alteglofsheim, Germany).
- buffer B: 0.1% TFA in acetonitrile filtered and degassed as for buffer A.

4.3.2
Step by Step Chromatography of Proteins and Peptides

If an separating an unknown sample mixture for the first time using this system follow these steps:

1. Equilibrate the column for 5 min.

2. Start the first run with a linear buffer gradient of 0%B to 100%B in 30 min at ambient temperature with a flow rate of 1 ml/min and detection wavelength of 220 nm **without sample application.**

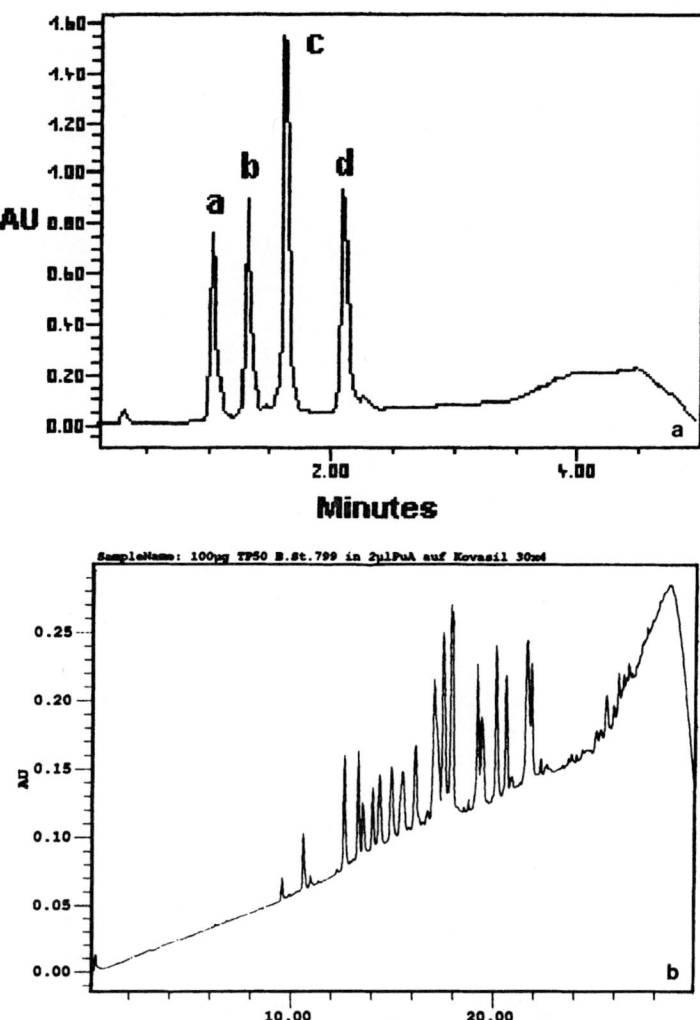

Fig. 1. a The separation of four proteins on a 33 x 4 mm Micra NPS 1.5μm column with a buffer B(see text) gradient ranging from 20% to 60% in 3 min, 60% to 100% in 0.5 min, 0.5 min at 100% buffer B, and 100% to 20% buffer B in 0.5 min; 10 μg of each protein were dissolved in a total of 2 μl starting buffer. *a*, ribonuclease; *b*, cytochrome c; *c*, lysozyme; *d*, β–lactoglobulin. **b** The separation of ribosomal proteins from the 50S subunit of *B. stearothermophiles 799* on a 30 x 4 mm Kovasil 1.5μm column with a buffer B gradient ranging from 10% to 50% in 25 min, 50% to 80% in 2 min, 1 min at 80% buffer B, 80% to 10% buffer B in 1 min and 1 min at 10% buffer B; 100μg of total protein were dissolved in a total of 2 μl starting buffer. For both separations the temperature was 60°C and the detection wavelength 220 nm

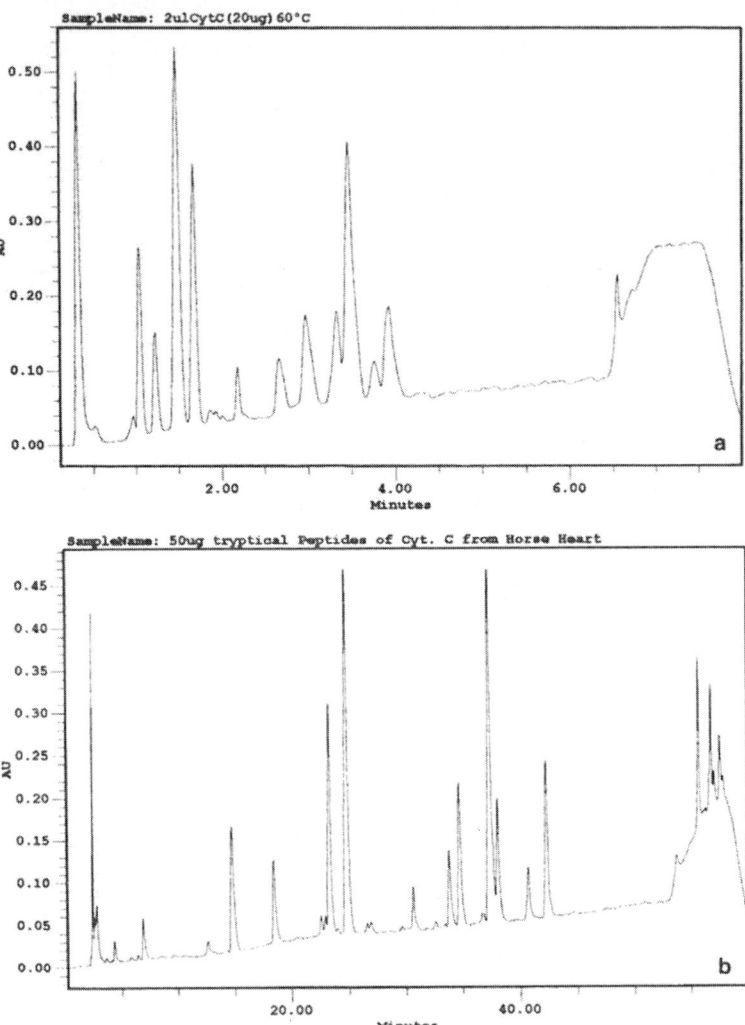

Fig. 2a,b. Comparison of the separation of tryptic peptides cytochrome c on two different columns. Cytochrome c was extracted from horse heart and enzymatically digested with trypsin for 4 h at ambient temperature. a Separation on the short column Kovasil MS–H; 20 µg of sample were directly injected. The buffer B gradient was: 10 to 30% in 6 min, 30% to 80% in 0.5 min, 0.5 min at 80% buffer B, 0.5 min from 80% to 10% buffer B and 0.5 min at 10%buffer B. b Separation on a 250 x 4.6 mm Vydac RP, C18 column with a particle diameter of 5 µm (filled by Knauer, Berlin); 50 µg were injected. The buffer B gradient was: 5% to 40% in 50 min, 40% to 80%in 2 min, 3 min at 80% buffer B, 2 min from 80% to 5% buffer B, and 3 min at 5%buffer B. For both separations, the flow rate was 1 ml/min, the column temperature 60 °C and detection wavelength 220 nm

3. If the run shows a stable baseline, apply 1–20 µg of proteins or peptides in 2–4 µl of HPLC-starting buffer A.

4. Optimize the gradient for sufficient resolution of all components by varying the gradient slope and the column temperature.

5. Collect the peaks for further investigations.

Figure 1 shows the separation of four proteins, ribonuclease, cytochrome c, lysozyme, and β-lactoglobulin. Figure 2 shows the separation of tryptic peptides from cytochrome c on two different columns.

4.4
Separation of Amino Acids

For the required HPLC equipment and the precolumn derivatization procedure, including detailed descriptions of the eluent and reagent compositions, see Chap. 15.

4.4.1
Materials

– 30 x 4 mm column filled with Eurospher C18, 1.5 µm non-   **Equipment**
  porous silica-based particles (Knauer, Berlin, Germany).
  Other nonporous columns from other companies may also be
  used.

– Promis autosampler with a precolumn derivatization unit
  (Spark Holland, Ve Emmen, The Netherlands).

4.4.2
Step by Step Chromatography of Amino Acids

1. Equlibrate the column with starting buffer for 5 min.

2. Start the above gradient **without injecting the sample.**

3. When the baseline is stable, reequilibrate the column again for 5 min.

4. Derivatize10–100 pmol of amino acid mixture in 10 µl sample   **Sample**
   buffer with 10 µl of OPA solution and inject it with the auto-   **preparation**
   sampler after exactly 1 min of incubation at room temperature. Start the gradient run

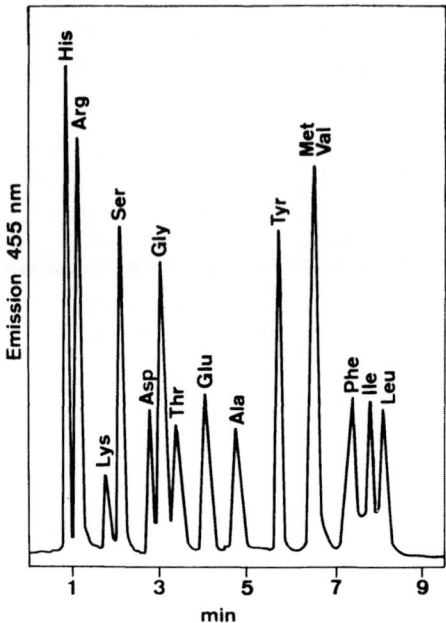

Fig. 3. The separation of 14 amino acids, 100 pmol each, after precolumn derivatization with o-phthaldialdehyde, within 7 min. The flow rate was 1.5 ml/min
and fluorescence detection took place with an excitation wavelength of 340 nm
and emission at 455 nm using fluorescence detector RF–551 from Shimadzu
(Kyoto, Japan). For buffer composition, see text. The buffer B gradient started
with 2% to 10% in 4.5 min, continued with 10% to 20% in 0.5 min, 20% to 30%
in 3 min and 30% to 2% in 2 min at a column temperature of 30 °C

5. Integrate the peaks and calculate the amino acid composition
    (see Chap. 15).

Figure 3 shows the separation of 14 amino acids, 100 pmol each,
after precolumn derivatization with o-phthaldialdehyde. Buffer A
consisted of 30 mM sodium hydrogen phosphate and 1 ml
tetrabutyl ammonium chloride brought to a final volume of 1 l
with deionized water and a pH of 5.2. Buffer B was HPLC grade
methanol.

4.5
Troubleshooting

By using nonporous silica-based short columns the complete
flow path volume, dwell volume, flow cell volume and injection
volume should be as low as possible. Increased resolution may be
achieved by reducing the flow path volume. While a flow rate of
1 ml/min or 1.5 ml/min is used the back pressure is always higher

due to the 1.5 μm columns but can decreased by increasing the column temperature. Higher temperature also increases the resolution, which is sometimes necessary for satisfactory separation. To protect the column, the use of filtered buffers and an inlet filter with a low path volume is highly recommended. Use of aqueous buffers with a salt concentration more than 10 mM can cause crystalization of salt at pump heads resulting in small leaks. This can be prevented by washing the pump head through the drain holes continuously with water and draining the complete system after use.

- Always use highly pure HPLC grade solvents and probes and an inlet filter.

Maintenance of columns

- Prior to use wash the column with 80% methanol followed by aqueous buffer for 15 min.

- Check column purity by a gradient run without sample application.

- Test the efficiency of the column by running a protein, peptide or amino acid standard.

- Never use buffers with a pH < 2 and > 8 because of silica hydrolysis.

- If back pressure increases, change the inlet filter.

- If "ghost" peaks appear, regenerate the column as follows:
  - Wash column with water to obtain stable baseline.
  - Replace water by methanol and wash again to obtain stable baseline.
  - Replace methanol by chloroform and wash out all impurities
  - Wash with methanol.
  - Equilibrate the column with starting aqueous buffer.
  - Check the resolution by injecting a standard mixture.

- Store the column in 80% methanol.

4.6
Comments

Nonporous silica-based particles improve the efficiency of chromatography columns. Use of microsphere particles allows for the rapid separation of large and small biomolecules such as proteins, peptides and amino acids. In addition, the separated biological molecules elute in narrow bands ranging from a few microliters to up to more than 1 ml. This avoids multiple concentration steps and reduces exposure to severe conditions. In spite of the smaller sample which passes the flow cell, the detection sensitivity is higher than with porous silica. Columns packed with nonporous material are strong enough to operate up to the limit of 300 bars, depending upon the column and flow rate used. New filter techniques have led to an enormous reduction in problems caused by clogging of the column flow path.

References

Janzen R, Unger KK, Giesche H, Kinkel JN, Hearn MTW (1987) Performance on nonporous monodisperse 1.5μm bonded silicas in the separation of proteins by hydrophobic interaction chromatography. J Chromat 397: 91–97

Jilge G, Janzen R, Giesche H, Unger KK, Kinkel JN, Hearn MTW (1987) Retention and selectivity of proteins and peptides an gradient elution on nonporous monodisperse 1.5μm reversed phase silicas. J Chromat 397: 71–80

Kalghatgi K, Horvath C (1987) Rapid analysis of proteins and peptides by reverse-phase chromatography. J Chromat 398: 335–339

Kalghatgi K, Horvath C (1989)In: Wittmann–Liebold B (ed) Rapid peptide mapping and protein analysis by HPLC. Methods in protein sequence analysis. Springer, Berlin, Heidelberg, New York, pp 248–255

Unger KK, Jilge G, Kinkel JN, Hearn MTW (1986) Evaluation of advanced silica particles for the separation of biomolecules by high performance liquid chromatography. J Chromat 359: 61–72

Unger KK, Jilge G, Janzen R, Giesche H, Kinkel JN (1986) Nonporous microparticulate support in high performance liquid chromatography (HPLC) of biopolymers: concepts, realizations and prospects. Chromatographia, 22: 379–380

Stromer D, Gottschall K, Kamp RM, (1993) Schnelle Trennung von Biomolekülen. BioTec 4: 32–37

Part II
Separation of Peptides

Enzymatic and Chemical Cleavages of Proteins

R. KRAFT

5.1
Background

The main goals in the characterization of proteins are: (1) to get enough amino acid sequence to identify a purified protein or an "interesting spot or band" on a gel, (2) to construct oligonucleotide probes for deduction of the primary structure by recombinant DNA techniques and/or (3) to prepare anti-peptide antibodies. Today, sequence information can be obtained routinely from less than 10 pmol of polypeptide if the NH_2-terminal is end is not blocked (Kraft et al. 1995). Unfortunately, about 80% of the intracellular soluble proteins from eukaryotic cells are known to have blocked NH_2-terminals as a result of posttranslational modifications and therefore are refractory to Edman degradation. Additionally, isolation procedures of microquantities of proteins can result in artificial blocking of their NH_2-terminals (Kraft et al. 1988; Bonaventura et al. 1994). Therefore sequence data can only be obtained by removing these blocking groups (see Chap. 12) and/or by internal cleavage of the respective proteins followed by sequencing of the single peptide chains separated and isolated from the resulting mixtures. Beyond this, determination of sequences from different peptides obtained from the cleaved protein leads to more accurate identification of the protein and allows for more alternatives in subsequent molecular cloning projects. Furthermore, at least some protein sequence data are necessary for interpretation of DNA sequence data, identification of protein modifications (acylation, methylation, glycosylation, phosphorylation, cleavage of signal or prosequences, etc.) and interpretation of protein crystallographic data. Taken together there is an urgent need in protein chemistry for efficient, sensitive and reliable methods to: (1) fragment purified proteins chemically or enzymatically, (2) separate the generated internal peptides from the complex mixtures and (3) identify isolated peptides by subsequent analytical procedures (amino acid analysis, automated sequencing or mass spectrometric methods).

5.2
Principle and Applications

The availability of an arsenal of chemical reagents (Han et al. 1983) and proteolytic enzymes (Flannery et al. 1989) with different cleavage specificities makes it possible to obtain overlapping peptides allowing more extended sequences to be deduced. Proteases can introduce restricted cleavages into a polypeptide substrate. Thus they are valuable tools to obtain higher order structural data about exposed loops within more protease resistant domain cores or about linking regions of polypeptide chains between domains (Price and Johnson 1989; Misselwitz et al. 1992). In addition, proteases are useful topological probes for membrane proteins (Pratt 1989; Chernogolov et al. 1994), and they provide a fast and efficient method for obtaining soluble fractions of membrane proteins in high yields (Butler 1989). Finally, limited proteolysis can be used to identify amino acids and peptide regions at the surface of protein complexes, as has recently been shown for intact ribosomal subunits (Kruft and Wittmann-Liebold 1991). In spite of the relevance of restricted cleavages for many research projects, in this chapter protocols that lead to more extensive digestion of denatured proteins are given. The aim of this approach is to obtain as many peptides in the highest yields possible for primary structural work.

The choice of the cleavage mode depends primarily on the number and size of fragments desired (Matsudaira 1993). In the case of chemical cleavage at less abundant residues (e.g., Met, Asp-Pro or Trp) SDS-PAGE (Tricine-gels according to Schägger and von Jagow 1987) with subsequent electroblotting of the single peptide bands onto an inert membrane (such as polyvinylidene difluoride, PVDF) seems to be the method of choice to prevent sample losses of microquantities of protein. By contrast, proteolytic digestion at high frequency residues (e.g., Lys or Arg) of small sample amounts may advantageously be resolved by the use of narrow bore column reverse-phase HPLC.

5.3
Chemical and Enzymatic Cleavage Reagents

The most frequently employed chemical procedure is cleavage of the polypeptide chain at the carboxylic side of methionine by cyanogen bromide (CNBr) in 70% formic or trifluoroacetic acid.

A major advantage of this method is the simple removal of the volatile is by products by vacuum concentration in a microfuge. Moreover, in the case of in situ cleavage of proteins contained in a gel or on a membrane, the 70% formic acid, a good solvent for polypeptides, facilitates extraction of the fragments generated from the gel matrix or from PVDF into the supernatant. If low peptide recovery is observed for blotted proteins (for example, very long and hydrophobic fragments may be retained on the support), it may be of advantage to subdigest the PVDF-remained fragments once more with a protease.

A survey of the enzymatic procedures for many applications described in the literature showed that trypsin was the enzyme of choice for in situ digestion of proteins; however, chymotrypsin, endoproteinases Lys-C and Asp-N work equally well (Fernandez et al. 1992). The availability of several proteases improves the chances of recovering optimally sized peptides. Fragments of about 15 amino acids are preferred for oligonucleotide probe design for cloning.

Little is known about the structural changes of proteins immobilized on supports (hydrophobic forces, ionic interactions), the resulting substrate/protease interactions or the kinetics of the reaction. Most commonly the digestions are performed at 37 °C for at least 15 h, often in the presence of limited amounts of detergents, denaturants or organic solvents in the cleavage buffer. As has recently been found (Fernandez et al. 1994) there is no need for separate pretreatment with a blocking reagent (e.g., PVP-40) to avoid adsorption of the enzymes to the membrane: Hydrogenated Triton X-100, which is included in the cleavage buffer as an efficient solubilizing reagent for peptides, also prevents enzyme binding to the membrane. At submicrogram levels it is essential to maximize protease concentration. In solution, for denatured proteins and for limited digestion under native conditions, enzyme to substrate ratios are usually 1/20 to 1/100. For immobilized substrates, however, a ratio of about 1/5 to 1/20 has been recommended. In the latter case the amount of support should be kept as limited as possible.

The cleavage procedures listed in Table 1 were found to be very useful. Of course many other reagents and procedures are also available for polypeptide cleavage, but those listed in Table 1 are a good choice for many situations.

Table 1. Cleavage procedures

Cleavage mode	Cleavage specificity	References
Chemical		
Cyanogen bromide	Carboxyl side of Met (particularly Met-Thr and Met-Ser give poor yields; sulfone or sulfoxide of Met are not cleaved at all)	Aitken et al. 1989[a] Jahnen et al. 1990[b] Bommer et al. 1991[b] Stone & Williams 1993[c]
BNPS-skatole	Carboxyl side of Trp	Aitken et al. 1989[a]
Dilute acid	Both sides of Asp Less frequently at Asn, Glu, Gln	Aitken et al. 1989[a] Vanfleteren et al. 1992[b]
Formic acid (70%)	Preferentially Asp-Pro bonds	Misselwitz et al. 1992[a] Sonderegger et al. 1982[b]
Hydroxylamine	Asn-Gly bonds	Aitken et al. 1989[a] Strydow & Kandror 1995[c]
Enzymatic		
Trypsin	Carboxyl side of Lys and Arg (exception is Lys-Pro, Arg-Pro to a lesser extent)	Rosenfeld et al. 1992[b] Fernandez et al. 1994[c]
Endopeptidase Lys-C	Carboxyl side of Lys (and S-aminoethylcysteine)	Jenö et al. 1995[b] Hellman et al. 1995[b]
Endopeptidase Arg-C	Carboxyl side of Arg	Fernandez et al. 1994[c]
Endopeptidase Glu-C (Protease V8)	Carboxyl side of Glu (in ammonium bicarbonate at pH 7.8, in ammonium acetate at pH 4.0, additionally at Asp in phosphate buffer at pH 7.8)	Fernandez et al. 1994[c] Sorensen et al., 1991[a] Jenö et al. 1995[b]
Endopeptidase Asp-N	Amino terminal side of Asp and at cysteic acid in case of oxidized proteins	Fernandez et al., 1994[c] Aitken et al. 1989[a] Jenö et al. 1995[b]
Chymotrypsin	Preferentially at the carboxyl side of Phe, Trp, Tyr and Leu	Aitken et al. 1989[a]
Thermolysin	Preferentially at the amino terminal side of hydrophobic residues (Leu, Ile, Phe, Val, Met, Trp)	Aitken et al. 1989[a]

[a] In solution.
[b] In gel.
[c] On support.

5.4
Methods

5.4.1
Sample Preparation

All polypeptides to be applied to sequence analysis must first be freed of buffer salts and other interfering compounds. For protein samples purified from HPLC or FPLC in milligram quantities of protein, it is relatively easy to achieve this by dialysis or gel filtration and to concentrate the protein by lyophilization or ultrafiltration. However, microgram quantities of material are considerably more difficult to handle because peptides and proteins at this level are especially susceptible to adsorptive losses. Possible procedures for separation and isolation of submicrogram amounts of proteins with highest resolution are one-dimensional (1-D) or two-dimensional (2-D) PAGE. There are several approaches for obtaining internal sequence data from proteins separated by SDS-PAGE: either one can digest the sample after electroelution from SDS gels (unfortunately, this often results in poor overall yields when working with picomole amounts), alternatively, one can digest the sample within a gel slice (Rosenfeld et al. 1992) or digest the sample after blotting onto membranes (Aebersold et al. 1987). Experimental details for cleavage procedures of gel or membrane immobilized proteins are given in Chaps. 7 and 8.

In general, proteins require denaturation and disulfide bond cleavage, before enzymatic digestion can go to completion. This can be accomplished with little or no loss of sample if the reducing and alkylating steps are performed in the SDS sample buffer just prior to electrophoresis (Jenö et al. 1995) or by in situ procedures directly on sequencer membrane supports (Ploug et al. 1992). Due to the additional steps involved in digestion and peptide isolation, these procedures typically require essentially more starting sample (about tenfold) than is needed for direct NH_2-terminal sequencing.

5.4.2
CNBr Cleavage

As mentioned above, cleavage at methionine residues with cyanogen bromide is the most frequently used chemical cleavage method. Cleavage occurs in high yields at the carboxyl side of methionine residues. Since methionine residues occur very rarely (average frequency of two per 100 amino acids) in proteins, usually a few large peptide fragments are generated suitable for electrophoretic separation. The reaction is usually carried out in 70% formic acid (or in 70% trifluoroacetic acid, favored in the case of Met-Ser and Met-Thr bonds) at room temperature for 24–48 h using a roughly 100-fold molar excess of CNBr over methionine residues present. Prior to CNBr cleavage it is advisable to reduce the protein, since methionine residues can be oxidized to sulfoxide or sulfone during isolation procedures. These oxidized residues are not cleaved by CNBr. As a result of the CNBr cleavage reaction all methionine residues are converted to homoserine or homoserine lactone. Therefore, only the peptide derived from the COOH-terminal of the molecule does not end in a homoserine residue.

1. Prepare a mixture of 70% formic acid by flushing with nitrogen gas (several minutes) to remove any soluted oxygen.

2. Dissolve 10 μg lyophilized protein in 90 μl of the nitrogen flushed formic acid.

3. Dissolve approximately 1 mg of CNBr in 200 μl 70% formic acid

   Note: CNBr crystals should be dry and colorless; handle this compound only under a fume hood because it is highly toxic and relatively volatile!.

4. Add 10 μl of the freshly prepared CNBr containing formic acid to the protein solution.

5. Keep the mixture in the dark under nitrogen at room temperature and incubate for 24 h.

6. Add Milli-Q water (5–10 volumes) and remove the excess reagent by vacuum concentration in a Speed Vac

   Note: great care should be taken to trap volatile cyanide products during the lyophilization step.

7. Redissolve the lyophilized CNBr fragments in electrophoresis buffer followed by SDS-PAGE or inject the peptides solubilized in solvent A (0.1% TFA/water) directly onto a reverse phase HPLC columnn.

5.4.3
Partial Acid Hydrolysis

Due to the relative lack of specificity partial acid hydrolysis is not a very popular method. Nonetheless, this cleavage procedure may be useful for samples when other fragmentation procedures have failed, particularly for proteins of limited solubility. In fact, some amino acids are more sensitive to diluted acids than other ones. Most sensitive is the Asp-Pro bond (average frequency is one bond per 400 amino acids). Moreover, cleavage at Asp and at other residues (for example, Asn, Glu, Gln, Ser, Thr) occurs, too, depending on the the particular protein and the conditions used (acid concentration, temperature and time). Therefore, the cleavage pattern is hard to predict. It should be noted that very harsh acid treatment of proteins can cause side reactions such as hydrolysis of amide groups, partial destruction of tryptophan or cyclization of NH_2-terminal glutamine residues.

1. Add to the protein, contained in a glass ampule, dilute acid (for example, 0.2% HCl, 2% formic acid, 2% acetic acid or 0.1% TFA). In the case of very unsoluble proteins dissolve the protein in more concentrated formic acid and then dilute.

2. Seal the ampule under nitrogen in vacuo and keep the sample for 2 h at 110 °C.

3. Open the ampule and dry in a Speed Vac. Because of the volatile reagents used no desalting step is necessary, and the mixture may be directly separated by HPLC or PAGE.

   Note: When specific cleavage at the Asp-Pro bond is desired, conditions such as 70% formic acid at 40 °C for 46 h may be successfully used. The cleavage is stopped by a 1:20 (v/v) dilution with water and the peptide material is lyophilized.

5.4.4
Trypsin, Lys-C and Arg-C

These enzymes produce fragments very suitable for sequence analysis followed by search in a database library because of their highly specific cleavage pattern, only COOH-terminal to basic amino acid residues. For complete digestion the substrates should be denatured previously, for example by heat treatment, addition of chaotropic salts (e.g. guanidine hydrochloride or urea) to the cleavage buffer (proteases retain their activity under mild denaturing conditions) or by reductive alkylation.

1. Dissolve the denatured protein at a concentration of 1–10 µg/µl in 0.05–0.1 M N-ethylmorpholine acetate (or ammonium bicarbonate) buffer usually at pH 8.0–8.5.

2. Add the enzyme in a minimal volume freshly prepared as recommended by the manufacturer to give a weight ratio of 1%–2% enzyme/protein.

3. Incubate with shaking for up to 4 h at 37 °C.

4. Stop the digest by adding 10% TFA or by freezing. Use digest (after removing of any unsoluble material by centrifugation) directly for HPLC mapping.

5.4.5
Chymotrypsin

This enzyme exhibits a much broader specificity than trypsin. Chymotrypsin preferentially hydrolyzes peptide bonds at the COOH-terminal side of aromatic amino acids such as Phe, Trp, Tyr and at Leu. Cleavage at Met, Gln and His is known to occur, too, with extended incubation periods.

Note: The protocol given for digestion with trypsin also works well for chymotrypsin but less cleavage time (1–3 h) is required. If a maximum number of small peptides is desired, 12–24 h periods may be used.

5.4.6
Thermolysin

Thermolysin has a broad specificity for the NH_2-terminal side of hydrophobic residues such as Leu, Ile, Phe, Val, Met, Trp. The NH_2-terminal sides of Ala, Tyr, Thr are cleaved more slowly. The enzyme is active in 6–8 M urea or 0.5%–1% SDS.

Note: Reaction conditions for trypsin are also suitable for thermolysin, but the buffer should contain calcium carbonate (1–5 mM) for enhancing the thermostability of this enzyme. Normally digestion is performed at 45 °C for 1 h.

5.4.7
Endopeptidase Glu-C (Protease V8)

This serine-endopeptidase cleaves peptide bonds COOH-terminal to Glu residues at a rate that is about 3000-fold faster than that for Asp residues (Sprensen et al. 1991). Ammonium bicarbonate buffer (pH 7.8) or acetate buffer (pH 4.0), recommended to restrict cleavage to Glu, only inhibit the enzyme.

Note: The best approach is to perform the cleavage in 0.05–0.01 M phosphate buffer with a reduced enzyme concentration and digestion time.

5.4.8
Endopeptidase Asp-N

Asp-N is a metalloprotease that cleaves the peptide bond NH_2-terminal to Asp and, in the case of an oxidized protein, at the cysteic acid too. To prevent additional cleavage at Glu, it is advisable to reduce the enzyme concentration and the incubation time.

Note: Use 0.05 M sodium phosphate (pH 8.0). Up to 1 M urea, guanidine hydrochloride or 0.01% SDS may be added in the case of proteins hard to solubilize. Incubate for 1–15 h at 37 °C.

References

Aebersold RH, Leavitt J, Savedra RA, Hood LE, Kent SBH (1987) Internal amino acid sequence analysis of proteins separated by one- or two-dimensional gel electrophoresis after in situ protease digestion on nitrocellulose. Proc Natl Acad Sci USA 84:6970–6974

Aitken A, Geisow MJ, Findlay JBC, Holmes C, Yarwood A (1989) Peptide preparation and characterization. In: Findlay JBC, Geisow MJ (eds) Protein sequencing – a practical approach, IRL, Oxford, New York, Tokyo, pp 43–67

Bommer UA, Kraft R, Kurzchalia TV, Price NT, Proud CG (1991) Amino acid sequence analysis of the β- and γ-subunits of eukaryotic initiation factor eIF-2. Identification of regions interacting with GTP. Biochim et Biophys Acta 1079:308–315

Bonaventura C, Bonaventura J, Stevens R, Millington D (1994) Acrylamide in polyacrylamide gels can modify proteins during electrophoresis. Anal Biochem 222:44–48

Butler PE (1989) Solubilization of membrane proteins by proteolysis. In: Beynon RJ, Bond JS (eds) Proteolytic enzymes – a practical approach, IRL, Oxford, New York, Tokyo, pp 193–200

Chernogolov A, Usanov S, Kraft R, Schwarz D (1994) Selective chemical modification of Cys^{264} with diiodofluorescein iodacetamide as a tool to study the membrane topology of cytochrome P450scc (CYP11A1). FEBS Lett 340:83–88

Fernandez J, DeMott M, Atherton D, Mische SM (1992) Internal protein sequence analysis: enzymatic digestion for less than 10 μg of protein bound to polyvinylidene difluoride or nitrocellulose membranes. Anal Biochem 201:255–264

Fernandez J, Andrews L, Mische SM (1994) An improved procedure for enzymatic digestion of polyvinylidene difluride-bound proteins for internal sequence analysis, Anal Biochem 218:112–117

Flannery AV, Beynon RJ, Bond JS (1989) Proteolysis of proteins for sequence analysis and peptide mapping. In: Beynon RJ, Bond JS (eds) Proteolytic enzymes – a practical approach. IRL Press, Oxford, New York, Tokyo, pp 145–161

Han KK, Richard C, Biserte G (1983) Current developments in chemical cleavage of proteins. Int J Biochem 15:875–884

Hellman U, Wernstedt C, Góñez J, Heldin C-H (1995) Improvement of an "in-gel" digestion procedure for the micropreparation of internal protein fragments for amino acid sequencing. Anal Biochem 224:451–455

Hirano H, Komatsu S, Kajiwari H, Takagi Y, Tsunasawa, S (1993) Microsequence analysis of the N-terminally blocked proteins immobilized on polyvinylidene difluoride membrane by western blotting. Electrophoresis 14:839–846

Jahnen W, Ward LD, Reid GE, Moritz RL, Simpson J (1990) Internal amino acid sequencing of proteins by in situ cyanogen bromide cleavage in polyacrylamide gels. Biochem Biophys Res Commun 166:139–145

Jenö P, Mini T, Moes S, Hintermann E, Horsz, M, (1995) Internal sequences from proteins digested in polyacrylamide gels. Anal Biochem 224:75–82

Kraft R, Fessel R, Büttner D, Etzold G (1988) Manual N-terminal microsequencing of proteins electroeluted from polyacrylamide gel slices. Biol Chem Hoppe-Seyler 369:87–91

Kraft R, Kostka S, Hartmann E (1995) Microsequencing of proteins of the rough endoplasmic reticulum (rER) membrane. In: Atassi MZ Apella E. (eds) Methods in protein structure analysis – 1994. Plenum Corporation New York, pp 445–454

Kruft V, Wittmann-Liebold B (1991) Determination of peptide regions on the surface of the eubacterial and archaebacterial ribosome by limited proteolytic digestion. Biochemistry 30:11781–11787

Matsudaira P (1993) A practical guide to protein and peptide purification for microsequencing. Academic, Inc USA

Misselwitz R, Kraft R, Kostka S, Fabian H, Welfle K, Pfeil W, Welfle H, Gerlach, D (1992) Limited proteolysis of streptokinase and properties of some fragments. Int J Biol Macromol 14:107–116

Ploug M, Stoffer B, Jensen AL (1992) In situ alkylation of cysteine residues in a hydrophobic membrane protein immobilized on polyvinylidene difluoride membranes by electroblotting prior to microsequence and amino acid analysis. Electrophoresis 13:148–153

Pratt JM (1989) Proteases as topological probes for membrane proteins In: Beynon RJ, Bond JS (eds) Proteolytic enzymes – a practical approach. IRL, Oxford, New York, Tokyo, pp 181–191, 1989

Price NC, Johnson CM (1989) Proteinases as probes of conformation of soluble proteins. In: Beynon RJ, Bond JS (eds) Proteolytic enzymes – a practical approach. IRL, Oxford, New York, Tokyo, pp 163–179, 1989

Rosenfeld J, Capdevielle J, Guillemot JC, Ferrara P(1992) In gel digestion of proteins for internal sequence analysis after one- or two-dimensional gel electrophoresis. Anal Biochem 203:173–179

Schägger H, von Jagow G (1987) Tricine-sodium dodecyl sulfate-polyacrylamide gel electrophoresis for the separation of proteins in the range from 1 to 100 kDa. Anal Biochem 166:368–379

Sonderegger P, Jaussi R, Gehring H, Brunschweiler K, Christen P (1982) Peptide mapping of protein bands from polyacrylamide gel electrophpresis by chemical cleavage in gel pieces and re-electrophoresis. Anal Biochem 122:298–301

Sørensen SB, Sørensen TL, Breddam K (1991) Fragmentation of proteins by *S. aureus* strain V8 protease. Ammonium bicarbonate strongly inhibits the enzyme but does not improve the selectivity for glutamic acid. FEBS Lett 294:195–197

Stone KL, Williams KR (1993) Enzymatic digestion of proteins and HPLC peptide isolation. In: Matsudaira P (ed) A practical guide to protein and peptide purification for microsequencing. Academic Press, Inc USA, pp 45–67

Strydom DJ, Kandror, K (1995) Identification of actin isoforms after *in situ* hydroxylamine cleavage on sequencer membranes. Serum actin is a cytoplasmic isoform. Biochem Biophys Res Commun 166:139–145

Vanfleteren RJ, Raymackers JG, Vanbun SM, Meheus LA (1992) Peptide mapping and microsequencing of proteins separated by SDS-PAGE after limited *in situ* acid hydrolysis. Biotechniques 12:550–557

Chapter 6

Separation of Peptides Using HPLC and TLC Fingerprints

T. CHOLI-PAPADOPOULOU and R. M. KAMP

6.1
Introduction

The development of new, high sensitivity methods in protein chemistry (microisolation, amino acid microanalysis and microsequencing) has permitted characterization of a large number of previously unknown polypeptides. Initial studies were done with small peptides and have progressed to medium-size polypeptides.

Moreover, less and less material is necessary for structural characterization by amino acid analysis and sequencing. Whereas several nanomole of purified material were previously necessary, picomole amounts are now required for partial or complete structural characterization. Over the years, reversed-phase liquid chromatography (RP-LC) has evolved as the dominant tool for isolation and characterization of polypeptides. However, alternative modes of chromatography, such as thin-layer chromatography, is also used for peptide isolation.

6.1.1
Thin-Layer Fingerprints

Peptides can be separated with good resolution by two-dimensional fingerprints (Wittman-Liebold and Kamp 1980; Wittman-Liebold and Lehmann 1980), although there are some disavantages concerning their low recovery rates – in the range of 40%–70% depending on the type of peptide and the solvent used for the elution. Peptides purified by thin-layer fingerprints are suitable for direct microsequence analysis without any additional purification steps. The first-dimension peptide separation is electrophoresis, which depends on the peptide's net charges and its molecular mass (Offord 1966). In the second dimension the peptides are separated by ascending chromatography depending on their individual distribution coefficient. Several buffers are used for both dimensions. Electrophoresis is always performed prior to chromatography to remove traces of salts, which disturb the fingerprinting and should thus be totally avoided. After about 2 h of electrophoresis with cooling, the cellulose sheets are dried at room temperature Never use high temperatures or a warm fan. The ascending chromatography needs about 6–7 h. The dry cellulose sheets are stained with 0.15% ninhydrin in collidine/acetic acid/ethanol (30:100:870 v/v). The peptide spots are scraped out and eluted with 50% acetic acid for basic peptides or 20% pyridine or 0.07% ammonia for acidic peptides.

6.1.2
HPLC Purification

Peptide mapping by HPLC is a powerful technique that is widely applied to the structural analysis of proteins. Current methods employ columns with 5–10 µm macroporous alkyl silica sorbents and gradient elution with increasing organic solvent concentration. Further developments of the technique were aimed at increasing speed and sensitivity (Kalghatgi and Horvath 1988; Lu and Lai 1986; Stone and Williams 1986; Wilson et al. 1986). The choice of reversed-phase supports depends on the application. The separation of small peptides (< 30 amino acids) can be performed on a C_{18} support of small pore size (< 100 Å) with a particle size of 5–10 mm and a column length of 10–25 cm. Examples of C_{18} supports with proven ability in small peptide fractionization are the Altex Ultrasphere C_8 and C_{18} (Beckmann, Munich, Germany) and the Waters NovaPak C_{18} (Waters, Milford, MA, USA) columns. Separations of larger peptides can also be performed on the Vydac C_{18} (The Separation Group, Hesperia, CA, USA) and Synchropak C_{18} 300 Å (Synchrom, Linden, USA) supports.

Peptide mapping and further sequence analysis is important for: (1) the construction of oligonucleotide probes for molecular cloning; (2) providing confirmatory evidence for cDNA-deduced amino acid sequences; (3) fingerprinting recombinant DNA-derived proteins; (4) complete protein structure determinations; (5) identifying posttranslational modification sites; (6) epitope mapping and (7) disulfide bond assignment. Peptide mapping of subnanomole quantities of protein causes a number of problems and requires additional sample preparation steps (desalting after reduction and alkylation, sample concentration, multistep chromatographic purification protocols) which can be omitted by using microbore chromatography. Small and highly charged peptides which are not retained on conventional RP-HPLC systems and elute with the column breakthrough can be isolated using small pore size support (e.g., ODS-Hypersil, 80 Å) and ion-pairing reagents in the mobile phase. Peptides purified in this way can be directly sequenced without the need for desalting.

6.1.3
Combined In Situ Digestion and HPLC Separation

If the NH_2-terminal of a protein is blocked, useful sequence information can be obtained after in situ CNBr treatment of the sample on the glass fibers (Simpson et al. 1984). In this case desalting of the peptide mixture by RP-HPLC short columns is necessary (Kamp 1986). Compared with the separation times of classical methods (dialysis, chloroform-methanol or trichloroacetic acid precipitation and open column chromatography), that of RP-HPLC is very short – between 5 and 10 min. Thus problems due to loss of material and solubility are minimized. The desalted and dried peptide mixture is applied directly into the sequencer without prior separation of the peptides. This strategy can be followed if the protein sequence is known and only information for structural studies is needed. In the case of membrane proteins this approach is quite difficult due to the hydrophobic character of the peptides. Thus, instead of RP-HPLC techniques, blotting onto PVDF membranes and sequencing are suggested in order to get internal sequence information. The produced peptides are separated on RP-HPLC, without desalting in the case of CNBr peptides, and subjected to Edman degradation. Direct NH_2-terminal sequence analysis of electroblotted proteins resolved by SDS-PAGE, first introduced by Vandekerckhove and Aebersold (Vandekerckhove et al. 1985; Aebersold et al. 1986), is now an important strategy for protein microsequencing. In situ enzymatic digestions or chemical cleavage with CNBr on PVDF membranes has also been described by Fimmel et al. (1989) and Choli et al. (1990). In the former, recovery of released peptides is improved by the addition of 10%–20% methanol to the digestion buffer without significant inhibition of enzymatic activity. The presence of even 10% acetonitrile in the digestion buffer dramatically decreases activity, regardless of the enzyme used. In any case treatment of the PVDF membranes is superior. The limited solubility of hydrophobic proteins, e.g., integral membrane proteins and membrane associated proteins/peptides, requires the presence of ionic detergents such as SDS during purification. Preparations of these proteins therefore contain various amounts of SDS which interfere (above 0.1%) with subsequent enzymatic fragmentation and peptide separation. The separation resolution of HPLC reversed-phase columns is remarkably reduced or even destroyed in the presence of a few micrograms of SDS. For removal of SDS from peptide mixtures the most effective solution was made up of heptane and isoamyl alcohol. Posttranslationally modified peptides can also be separated and identified by HPLC. Peptides with the same amino acid composition, except for lysine content, elute at different times (Choli et al. 1988a,b). This effect is probably caused by the different degree of methylation of lysine residues.

6.2
Separation of Peptides Using HPLC

HPLC is a very useful technique for the separation of peptides. Volatile and UV-transparent buffers offer the advantages of high sensitivity runs and direct microsequencing, electrophoresis or amino acid analysis without any additional desalting steps.

Depending on the peptide mixture, three different volatile buffer systems can be used:

- System 1: buffer A (0.1% trifluoroacetic acid, TFA) and buffer B (acetonitrile with 0.1% TFA)

- System 2: buffer A (ammonium formate pH 7.8 prepared from 1.5 ml 25% ammonia and 0.25 ml 98% formic acid/2 l water) and buffer B (methanol with 20% buffer A)

- System 3: buffer A (ammonium formate pH 4.4 prepared from 0.4 ml 25% ammonia and 0.25 ml 98% formic acid/2 l water) and buffer B (methanol with 20% buffer A)
  As an example the separation of tryptic peptides of cytochrome c will be discussed.

Materials
- HPLC system consisting of two pumps, mixing chamber, injection valve or autosampler, UV detector (220 nm) or diode array detector from Waters (Milford, MA, USA)

- HPLC column, e.g., Vydac C_{18} column (250 x 4 mm, pore size 300 Å, 5 µm particle size) and guard column

- redistilled trifluoroacetic acid (TFA)

- acetonitrile HPLC grade

- HPLC water

- column oven

- syringes

1. Equilibrate the column for 20 min.   HPLC separation

2. Start the gradient, **without injecting any sample**, to test the purity of the column.

3. If the run shows a stable baseline and "ghost" peaks do not appear, inject the sample.

4. Inject 100 µg tryptic peptide mixture in 100 µl 0.1% TFA after short centrifugation to remove insoluble particles.

5. Set the UV detector on 220 nm and program a flow rate of   Separation
   1 ml/min. The temperature during separation is 25 °C. Start   conditions
   the gradient as follows:

   – 5%B to 40%B in 50 min

   – 40%B to 80%B in 2 min

   – hold at 80%B for 3 min

   – 80%B to 5%B in 3 min
     The separation of tryptic peptides is shown in Fig. 1.

6. Collect the fractions and dry the peptide containing fractions using, e.g., lyopilization (SpeedVac concentrator). Dissolve the dried sample in 50 µl 0.1% TFA or 50% pyridine/water and transfer to an amino acid analysis tube or onto PVDF membrane for sequencing.

A typical HPLC column for separation of peptides is C_{18} alkylated   Comments
silica. Depending on the hydrophobicity a C4-, C8- or C18 alky-

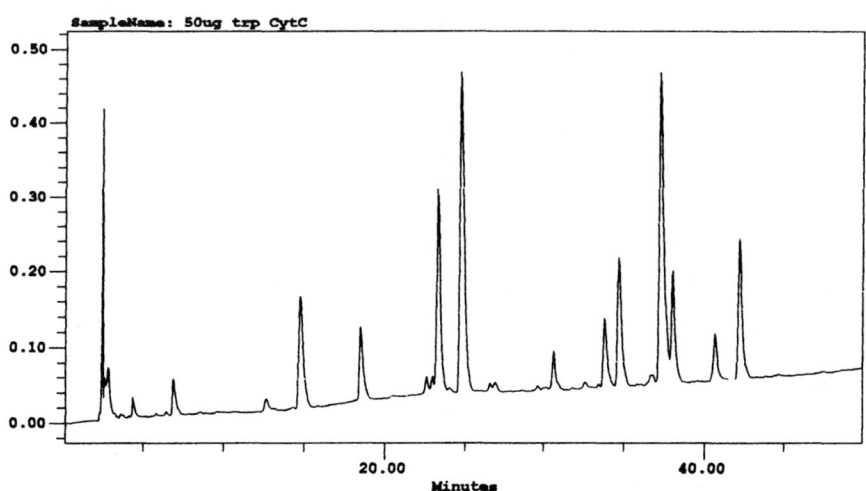

Fig. 1. Separation of cytochrome c tryptic peptides on Vydac C_{18} column. For separation conditions see Sect. 6.2 in text

lated support can be used. When separating hydrophobic and larger peptides C4 columns are more suitable, whereas for separation of hydropilic and smaller peptides C18 columns are recommended. For good recovery of peptides the pore size of the support can be selected according to the length of the peptides. Short hydrophilic peptides should be separated on 80 Å support, larger peptides are obtainable in better yields on 300 Å. Good reproducibility from run to run can be achieved using spherically shaped silica. The use of an irregular support and material from different batches or suppliers causes problem with satisfactory reproducibility. HPLC allows very quick and excellent separation of hydrophobic peptides and high resolution. The amounts necessary for sequencing peptides separated by HPLC are approximately five times less than those necessary after purification with thin-layer fingerprinting.

6.3
Thin-Layer Fingerprints

The peptides can be separated on thin-layer by two-dimensional fingerprinting. The first dimension separates peptides according to their net charge. In the second dimension the peptides are

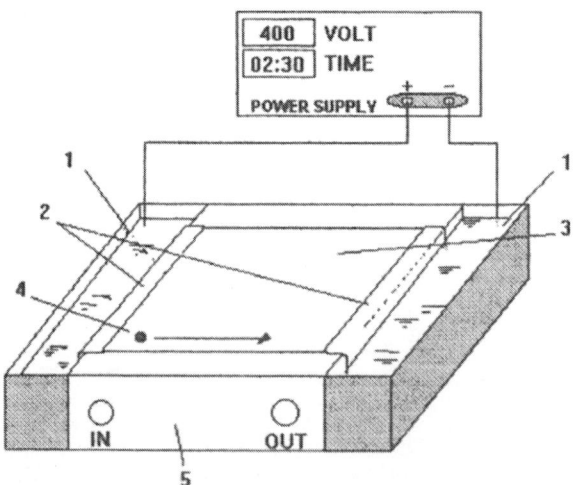

Fig. 2. Scheme of thin-layer electrophoresis chamber: *1*, buffer reservoir; *2*, electrode strip; *3*, thin-layer sheet; *4*, position of sample application; *5*, cooling plate

separated by chromatography depending on their hydrophobicity.

Materials

- thin-layer Cel 300 sheets (20 x 20 cm; 0.1 cm thick, Macherey & Nagel, Düren, Germany)
- thin-layer electrophoresis chamber with water cooling (Desaga, Heidelberg, Germany) (Fig. 2)
- 1–10 ml pipettes
- chromatography chamber for 20 x 20 cm sheets to perform ascending chromatography
- filter paper
- sharpened spatula

Buffers and solutions

- 50% acetic acid
- 50% formic acid
- 20% pyridine
- electrophoresis buffer: pyridine/acetic acid/acetone/water (50:100:375:1975, v/v)
- chromatography buffer: pyridine/butanol/acetic acid/water (50:75:15:60, v/v)
- ninhydrin solution: 0.1% ninhydrin in collidine/acetic acid/ethanol (30:100:870, v/v)
- fluorescamine solution: 0.05% fluorescamine in acetone and 5% pyridine in acetone

6.3.1
First-Dimension Electrophoresis

1. Dip the sheet into electrophoresis buffer, dry quickly with filter paper and place into electrophoresis chamber.

2. Open water cooling to the electrophoresis chamber (Fig. 2)

3. Fill the electrode reservoirs with electrophoresis buffer (pH 4.4); connect the thin-layer with buffer reservoir using electrophoresis strips.

4. Proof wetness grade of thin-layer by prerunning at 400 V for 5 min. Optimal conditions are: 12–18 mA for Cel 300 or 10–15 mA for Cel 400 sheets.

5. Apply 1–3 µl of 5 nmol tryptic peptides at the corner of the sheet (4 cm from the left side and 3 cm from the bottom). Do not touch the sheet with hands or pipette.

6. Add one drop DNP-OH (dinitrophenylsulfonic acid) or amido black as marker at the top of the sheet above the sample.

7. Close the electrophoresis chamber and start the electrophoresis at 400 V for 2 h with cooling.

8. Dry cellulose sheet after electrophoresis for 1 h at room temperature. Never use higher temperatures or a warm fan

6.3.2
Second-Dimension Chromatography

1. Place the cellulose sheet in a chromatography chamber for 7 h using PBEW buffer (dip the bottom only 5 mm into the buffer).

2. After chromatography dry the sheet for 1 h at room temperature.

3. Wet cellulose sheet with 5% pyridine solution in acetone. After 1 min drying at room temperature in the exhaust hood, wet thin-layer with 0.05% fluorescamine in acetone.

4. Dry the cellulose sheet for 15 min at room temperature.

5. Mark the peptides under UV light at 366 nm with soft pencil and scrape out.

6. Spray the cellulose sheet with 0.3% ninhydrin to detect peptide spots weakly staining with fluorescamine. Keep the sheet in the dark for complete development of spots.

7. Scrape out all peptides.

6.3.3
Elution of Peptides

The peptides from three thin-layer sheets are eluted and used for:

- amino acid analysis after OPA analysis:1/20 of amounts recovered

- end group analysis: 1/20 of amounts recovered

- sequencing: remainder

1. Scrape out peptide spots with sharpened spatula and transfer to Eppendorf tube.

2. Elute the peptides with acid or pyridine, depending on application:

   - 2 x 200 μl of 50% acetic acid for basic peptides

   - 2 x 200 μl of 20% pyridine for acidic peptides

   - 2 x 200 μl of 5.7 N HCl for peptides used only for amino acid analysis

   - 2 x 100 μl of 70% formic acid for peptides used for oxidation with performic acid (e.g., cysteine determination)

3. Stir scraped peptides carefully with vortexing and shake for 30 min using Eppendorf mixer.

4. Centrifuge the suspended cellulose 5 min at 5000 rpm and collect the supernatant into a hydrolysis or sequencing tube.

5. Repeat the elution of peptides for a second time and pool with the first extract.

6. Dry the peptides under vacuum and store at –20 °C.

6.3.4
Comments

For the first run the peptide mixture should be applied in the middle of the sheet, 3 cm from the bottom, and electrophoresed for 1 h at 400 V. After the first detection with ninhydrin, the next sample is placed near the cathode (usually 4 cm from the left side) for negatively charged peptides or near the anode for positively charged peptides.

The detection of peptides should be performed first with fluorescamine and subsequently with ninhydrin. The reaction with fluorescamine is reversible and the NH_2-terminal amino acid can be used for futher protein sequence analysis. The detection with ninhydrin is irreversible and destroys the NH_2-terminal amino group. Only a few end groups are suitable for futher sequencing. To omit this problem 0.15% ninhydrin, instead 0.3%, should be applied. The NH_2-terminal amino acids and lysine side chains are only partially destroyed after using diluted ninhydrin solution. Depending on the NH_2-terminal amino acids, some peptides give a specific color. NH_2-terminal glycine, threonine and serine peptides stain yellow, while NH_2-terminal tyrosine and histidine give a brown color. After development of the separated peptides with fluorescamine, quantitative amino acid analysis or sequencing is possible. It is possible to use special spray tests to detect the presence of arginine, tyrosine, and tryptophan. After spraying with 1% p-dimethylaminobenzaldehyde in 2 N HCl in acetone, tryptophan containing peptides result in red spots. Tyrosine containing peptides form red spots on a yellow bacground after spraying with 0.1% a-nitroso-b-naphthol in ethanol and after drying with 10% HNO_3.

Arginine containing peptides are visible under UV (254 nm) as yellow spots after spraying with 0.02% phenanthrene quinone in ethanol and 10% NaOH in 60% ethanol (v/v 1:1). The advantage of thin-layer fingerprinting is good reproducibility and resolution and easy preparation of peptides suitable for microseqence analysis. The disadvantage is the low recovery (40%–70%), depending on the type of peptide and the solvent for the elution. The use of fingerprinting techniques depends on the properties of the peptide mixtures and the equipment in the laboratory.

References

Aebersold RH, Teplow DB, Hood LE, Kent SBH (1986) Electroblotting onto activated glass. Biol Chem 261: 4229–4238

Choli T, Hennig P, Wittmann-Liebold B, Reinhardt R (1988a) Isolation, charakterization and microsequence analysis of a small basic methylated DNA-binding protein from the Archaebacterium Sulfolobus solfataricus. BBA 950: 193–203

Choli T, Wittmann-Liebold B, Reinhardt R (1988b) Microsequence analysis of DNA-binding proteins 7a, 7b and 7e from the Archaebacterium Sulfolobus acidocaldarius. J. Biol. Chem. 263: 7087–7093

Choli T, Kapp U, Wittmann-Liebold B (1990) Blotting of proteins onto Immobilon membranes. J Chrom 476: 59–72

Fimmel S, Choli T., Dencher NA, Buldt G, Wittmann-Liebold B (1989) Topography of surface-exposed amino acids in the membrane protein bacteriorhodopsin determineted by proteolysis and micro-sequencing. Biochim Biophys Acta 978: 231–240

Kalghatgi K, Horvath C (1988) Rapid peptide mapping by HPLC. J Chrom 443:343–354

Kamp RM (1986) High performance liquid chromatography of proteins. In: Wittmann-Liebold B, Salnikow J, Erdmann V (eds) Advanced methods in protein microsequence analysis. Springer, Berlin, Heidelberg, New York, pp 21–33

Lu H-S, Lai P-H (1986) Use of narrow-bore high performance liquid chromatography for microanalysis of protein structure. J Chrom 368: 215–231

Offord RE (1966) Electrophoretic mobilities of peptides on paper and their use in the determination of amide groups. Nature 211: 591–593

Simpson RJ, Nice EC (1984) In situ cyanogen bromide cleavage of N-terminally blocked proteins in a gas-phase sequencer. Biochem Int 8: 787–791

Stone KL, Williams KR (1986) High performance liquid chromatographic peptide mapping and amino acid analysis in the subnanomol range. J Chrom 359: 203–212

Vandekerckhove J, Bauw G, Puype M, Van Damm J, Van Montagu M (1985) Protein blotting on polybrene-coated glass-fiber sheets. Eur J Biochem 152:9–19

Wilson KJ, Huang AL, Brasseur MM, Yuan PN (1986) Microbore high performance liquid chromatography: An example of its applications in peptide characterization. Biochromatography 1: 106–112

Wittmann-Liebold B, Kamp RM (1980) Primary structure of ribosomal protein L29. Biochem Int 1: 436–445

Wittmann-Liebold B, Lehmann A (1980) New approaches to sequencing by micro-and automatic solid phase technique. In: Birr Ch (ed) Methods in peptide and protein sequence analysis. Elsevier, Amsterdam, pp 49–72

Chapter 7

Microseparation of In Situ Digested Peptides

A. OTTO

7.1
Introduction

High performance two-dimensional polyacrylamide electropho-
resis (HP-2-DE) allows separation of up to 10|000 proteins (Klose
1975; Klose and Kobalz 1995). After blotting the proteins from
the gels onto hydrophobic membranes NH_2-terminal sequence
information of blotted protein spots can be routinely obtained
from less than 10 pmols, if the NH_2-terminal is not blocked. Since
many proteins are NH_2-terminally blocked, it is necessary to generate fragments for obtaining internal sequence information. A
new strategy is shown in Fig. 1. The fragmentation of proteins in
situ in the polyacrylamide matrix is shown (left side, A). Among
the many methods published in the last years (Rosenfeld et al.
1992; Jenö et al. 1995; Hellman et al. 1995) we use that of Eckerskorn and Lottspeich (1989) and Kellermann et al. (1995) with
modification of the sample preparation prior to micro-HPLC
(peptide collecting device, Otto et al. 1996). The main advantage
of this new approach is the combination of elution and concentration of peptides after enzymatic digestion by a small amount
of reversed phase material. This procedure is simpler and more
effective than concentrating the protein from a number of gel
spots prior to enzymatic cleavage. The new peptide collecting
device (Fig. 2) described below does not require expensive
equipment and can be used easily in the laboratory. This approach is very useful in preparing probes for matrix-assisted
laser desorption/ionization mass spectrometry (MALDI-MS) of
peptide mixtures. Probes prepared by this method give excellent
mass spectra because buffer salts which might disturb the analysis are removed. In practice only 5% of the prepared enzymatic
digest has to be withdrawn for MALDI-MS analysis.

The in situ cleavage of proteins by enzymes blotted onto nitrocellulose (Aebersold et al. 1987) or PVDF membranes
(Madsudeira 1987) is an alternative tool for generating internal

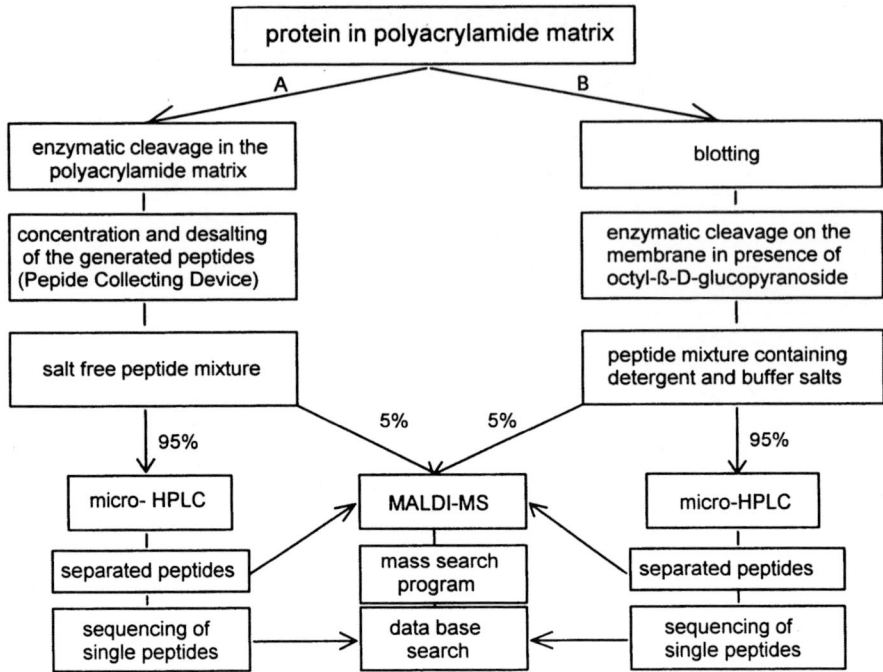

Fig. 1. Schematic flow chart demonstrating the strategy for the microseparation of a peptide mixture generated by in situ digestion of proteins after separation in polyacrylamide gels. Two ways are possible. *A*, in-gel digestion or *B*, digestion on membrane. MALDI-MS is recommended as the first step in the identification of proteins from internal peptide fragments

peptides (see B, right, in Fig. 1). As recently found (Fernandez et al. 1992, 1994), it is possible to get internal sequence information with amounts of less than 10 μg bound to the membranes by using digestion buffers containing a detergent such as hydrogenated Triton X-100 (Jungblut et al. 1994; Zeindl-Eberhart et al. 1994). The modification of this method with octyl-β-D-glucopyranoside as detergent was chosen because this detergent does not interfere with the MALDI-MS measurement. Such measurements can be used to obtain additional sequence information for identification of proteins listed in databases (Pappin et al. 1993; Thiede et al. 1996).

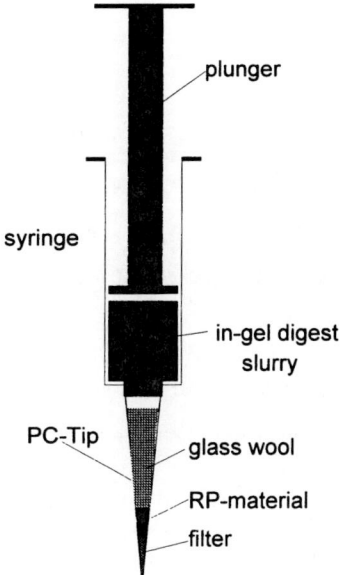

plunger

syringe

in-gel digest
slurry

PC-Tip

glass wool

RP-material

filter

Fig. 2. Peptide collecting device for concentration and desalting of peptides generated by in-gel digestion of proteins. For working procedure, see text

7.2
In Situ In-Gel Digestion of Protein Spots from HP-2-DE

7.2.1
Materials

- thermomixer (Eppendorf type 5436, Hamburg, Germany) **Equipment**
- 1 ml tuberculin syringes with stainless steel sieves at the bottom (WITA, Teltow, Germany)
- 1 ml tuberculin syringes (B. Braun, Melsungen, Germany)
- 2 ml disposable syringes (B. Braun, Melsungen, Germany)
- peptide collecting device (WITA, Teltow, Germany)
- microfuge (Beckman, Glenrothes, Scotland)
- SpeedVac (SAVANT, Farmingdale, NY, USA)
- Wash buffer: 100 mM Tris-HCl (pH 8.5) in 50% acetonitrile **Buffers**
- Digestion buffer 1: 100 mM Tris-HCl (pH 8.1), 1 mM $CaCl_2$, **and reagents** 10% acetonitrile
- Solution 1: 2% aqueous trifluoroacetic acid (TFA)
- Solution 2: 60% aceonitrile in 0.1% TFA
- Solution 3: 0.1% TFA in HPLC grade water

– trypsin (sequencing grade, Promega, Madison, WI, USA)

7.2.2
Procedure

Preparation of gel pieces

1. Cut out the stained protein spots from the gel and wash the gel pieces twice with wash buffer for 20 min at 30 °C (the buffer should cover the gel pieces).

2. Remove the supernatant from the shrunken gel pieces in both washes.

3. Add digestion buffer and let the gel pieces swell to original size, then remove the surplus buffer from the gel pieces.

4. Fill the gel pieces into an 1 ml tuberculin syringe without plunger (Luer connector cut off). This syringe is adapted with a stainless steel sieve (mesh size about 100 µm) at the bottom. Connect the plunger and press the gel through the sieve into an Eppendorf vial. This type of preparation of gel pieces is recommended for big gel pieces or gel bands, and if several gel pieces have to be digested. In case there are only two small pieces they may be shredded directly with a scalpel.

5. Shrink the gel material to approximately 50% in a SpeedVac (do not allow to dry completely).

Enzymatic digest

1. Swell the shrunken gel material in about 50–100 µl digestion buffer 1 containing 0.5–2 µg trypsin (enzyme:protein is 1:10).

2. Add approximately 100–300 µl of buffer (depending on the amount of gel), so that the gel forms a slurry; incubate at 37 °C for 15 h.

3. After digestion incubate with 2 volumes solution 1 (corresponding to the volume of digestion buffer 1) at 60 °C for 1 h with shaking.

Extraction of peptides

For the extraction of peptides from the gel slurry, a peptide collecting device is used (Fig. 2). This consists of a 2 ml disposable syringe and a specially prepared Eppendorf tip which is connected directly to the syringe. The peptide collecting tip (PC-tip) contains 5 mg of a selected reversed phase material and glass wool to fill the space.

Preparing the PC-tip:

1. Connect the PC-tip to a syringe filled with solution 2 and press 1 ml through the tip.

2. Disconnect the PC-tip and connect it to a second syringe filled with solution 3; press 1 ml through the tip.

3. Disconnect the PC-tip from the second syringe, fill up the tip with solution 3, and connect to an empty syringe without plunger.

4. Fill the gel slurry from the enzymatic digest after step 3 into the empty syringe with the aid of an 1 ml Eppendorf pipette (tip cut 2 mm below the top). Connect the plunger and press the gel slurry slowly through the PC-tip. Then immerse the gel in solution 3 and repeat the procedure.

5. Disconnect the syringe from the PC-tip and withdraw a part of the upper glass wool plug containing gel particles from the PC-tip with the aid of barbed tweezers.

6. Wash with 1 ml of solution 3 in the manner described above.

7. Take a clean 1 ml syringe filled with solution 2, connect the PC-tip and elute the peptide with 500 µl from the reversed phase material into an 1.5 ml Eppendorf vial.

8. Save a 5% aliquot for MALDI-MS analysis; lyophilize.

9. Remove acetonitrile from the remaining 95% in a SpeedVac and reduce the volume to about 60 µl; the sample is then ready for separation of peptides by micro-HPLC.

7.3
In Situ Cleavage of Proteins Blotted onto PVDF Membranes

7.3.1
Materials

- thermomixer (Eppendorf type 5436, Hamburg, Germany)   **Equipment**

- microfuge (Beckman, Glenrothes, Scotland)

- SpeedVac (SAVANT, Farmindale, NY, USA)

- wash solution: 50% acetonitriler   **Buffers**

- digestion buffer 2: 50 mM Tris-HCl (pH 8.5), 1 mM EDTA, 1%   **and reagents**
octyl β-D-glycopyranoside, 10% acetonitrile

- digestion buffer 3: 50 mM Tris-HCl (pH 8.1), 1 mM CaCl$_2$ 10% acetonitrile, 1% octyl β-D-glucopyranoside

- solution 2: 60% aceonitrile in 0.1% TFA

- solution 3: 0.1% TFA in HPLC-grade water

- solution 4: 10% aqueous TFA

- endoproteinase Lys-C (sequencing grade, Boehringer, Mannheim, Germany)

- trypsin (sequencing grade, Promega, Madison, WI, USA)

7.3.2
Procedure

Preparation of membrane pieces

1. Wash the protein spots blotted onto the membrane (protein stained with Coomassie blue or sulforhodamine) with wash solution for 1 min.

2. Cut the wet membrane pieces into approximately 1–2 mm squares and transfer the pieces into an 0.5 ml Eppendorf vial.

Lys-C digest

3a. Incubate with 10–30 µl digestion buffer 2 containing 0.2–0.5 µg endoproteinase Lys-C at 36 °C for 15 h (the buffer should cover the membrane pieces).

Trypsin digest

3b. Incubate with 10–30 µl digestion buffer 3 containing 0.2–0.5 µg modified trypsin at 37 °C for 15 h.

Extraction of peptides

1. Following digestion sonicate for 5 min, centrifuge, and transfer the supernatant into a new 0.5 ml Eppendorf vial.

2. Carry out consecutive washes with 20–50 µl each of solution 3 and 20–50 µl solution 2 with sonification and centrifugation; pool the fractions.

3. Use a 5% aliquot for MALDI-MS investigation and identification of the peptides by search in a mass database.

4. Remove the organic solvent from the remaining 95% in a SpeedVac by reducing the volume to approximately one fourth.

5. Acidify with 5µl 10% TFA.

7.4
Separation of the Cleavage Fragments

Peptides generated by enzymatic cleavage are separated on a micro-bore or narrow-bore reversed-phase liquid chromatography column. For optimal usage of such columns HPLC instruments should be available which can deliver stable flow rates from 20 µl/min to 200 µl/min and reproducible gradients at these flow rates. The detector cell should be adjusted to the small flow rates and should be coupled directly to the output of the HPLC column. The outlet capillary from the detector to the fraction collector unit should have a small inner diameter with a few microliters of content only to avoid peak broadening and a long delay time of the peptides collected in the tubes of the fraction collector. Peptides are detected at the leading wavelength of 214 nm. Two secondary wavelengths at 260 and 280 nm are desirable for the detection of aromatic or modified amino acids (e.g., pyridyl-ethylated cysteine).

7.4.1
Materials

- instrument suitable for micro-HPLC (SMART system, Pharmacia Biotech, Freiburg, Germany)   **Equipment**

- HPLC-column: μRPC C2/C18 SC2.1/10 (2.1 mm I.D., 10 cm length, 3 μm particle size, Pharmacia Biotech, Freiburg, Germany)

- SpeedVac (SAVANT, Farmingdale, NY, USA)

- microfuge (Beckman, Glenrothes, Scotland)

- 0.5 ml Eppendorf vials

- eluent A: 0.1% TFA in HPLC-grade water   **Solutions**

- eluent B: 0.095% TFA in acetonitrile

1. Centrifuge the 95% remaining peptide mixtures from either the in-gel digestion or from the blot digestion for 3 min in a microfuge at high speed.   **Conditions for microseparation**

2. Use the supernatant for application into the HPLC sample injector.

3. Carry out peptide separation with a linear gradient of acetonitrile in 0.1% TFA. The gradient is as follows:

   - 5% eluent B for 5 min

   - 5%–45% B in 5–85 min

   - 45–85% B in 85–105 min

   The flow rate is 100 μl/min and peak detection is carried out using a μPeak Monitor (Pharmacia Biotech, Freiburg, Germany), set at 214, 260 and 280 nm with automated peak collection based on the 214 nm wavelength.

7.5
Results and Comments

The results of the in situ digestion of proteins in the gel matrix or on membrane after blotting onto PVDF are shown in Figs. 3 and 4 as examples. Human heart proteins were separated in a series of 20 identical HP-2-DE gels. From a series of 10 gels a protein spot of interest, which was stained with Coomassie blue in the usual manner, was cut out. Ten spots were cleaved in the gel matrix by trypsin and the released peptides concentrated and desalted with the aid of the peptide collecting device according to pathway A in the flow chart in Fig. 1. An aliquot of 5% was taken for MALDI-MS analysis and 0.8% thereof used for sample preparation on the MS target. The resulting mass spectrum is shown in Fig. 5.

The main part (95%) of the purified peptide mixture was separated on a µRPC C2/C18 SC 2.1/10 column using the SMART system. Of the second series, eight gels were blotted onto PVDF, and after staining with sulforhodamine the protein spot of inter-

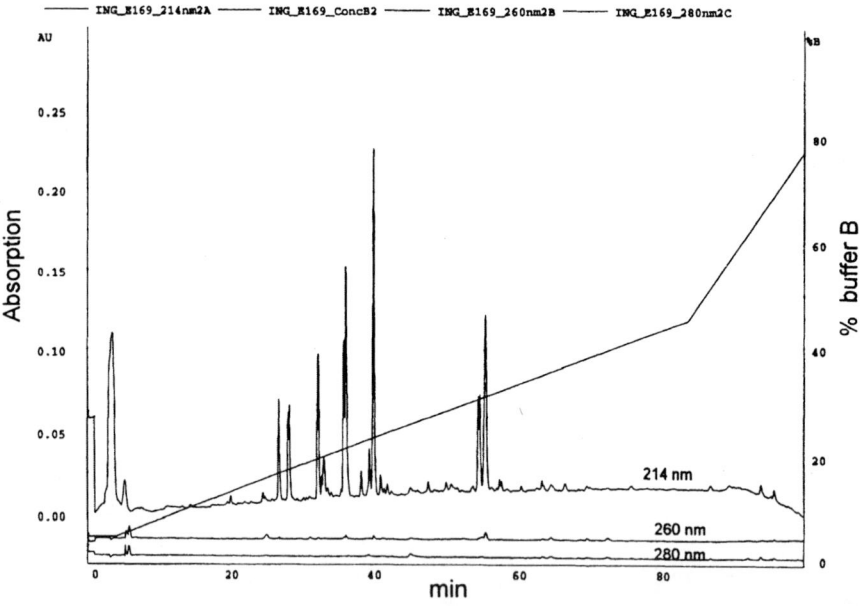

Fig. 3. HPLC peptide map of in situ in-gel digest: A tryptic digest was separated on a µRPC C2/C18 SC 2.1/10 column. Ten gel spots (from ten gels) derived from a human heart protein and separated by HR-2-DE were cleaved. The eluted peptides were concentrated and desalted by the peptide collecting device prior to injection onto micro-HPLC column. The protein was identified as myosin ligth chain 2 by MALDI-MS (see Fig. 5)

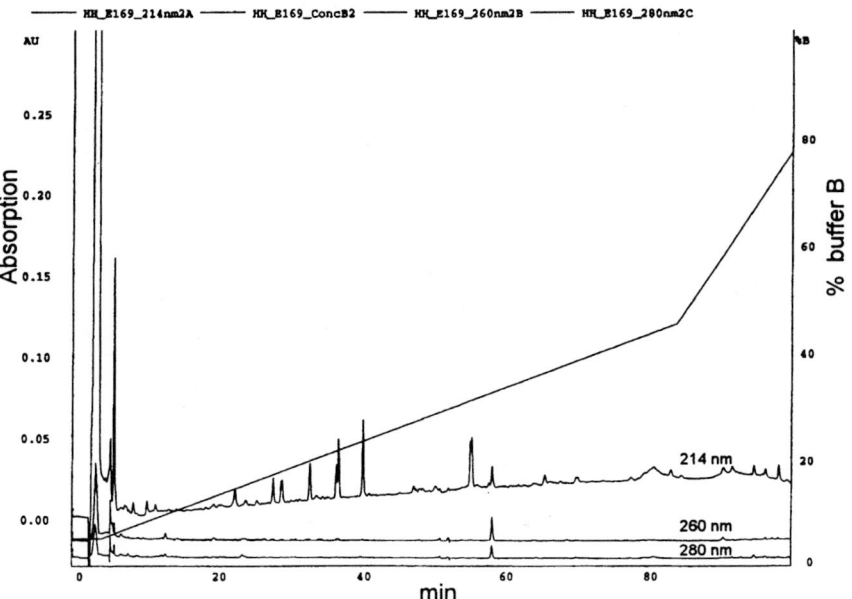

Fig. 4. HPLC peptide map of in situ membrane digest: Peptides from a tryptic digest were separated using a μRPC C2/C18 SC 2.1/10 column. Spots from eight HP-2-DE gels taken after blotting onto PVDF membrane were used. The protein was the same as shown in Fig. 3

est (same position as in the first series) was cut out from the membranes. These were digested with trypsin in the presence of the detergent octyl-β D-glucopyranoside according to pathway B in the flow chart shown in Fig. 1. Then, 5% was again taken for MALDI-MS-analysis and 0.8% thereof used for MS sample preparation. The remaining 95% of the peptides released from the membrane were separated in the same manner as described above. The in-gel digestion and the digestion on the PVDF membrane gave comparable peptide maps as shown in Figs. 3 and 4, but the yield of released peptides in the case of the in-gel digestion was about three times higher than obtained following in situ cleavage on the membrane (taking into account that only eight spots prior to cleavage on membrane were used).

MALDI-MS analysis was used in combination with the mass search program FRAGMOD (Thiede et al. 1996). Mass analysis unequivocally identified the protein in both preparation methods as myosin light chain 2. The protein could not be identified using a search in the MOWSE peptide mass database (Pappin et al. 1993), because this program does not consider modifications such as oxidation at methionine. The MALDI-MS spectrum of the peptide mixture prepared by the peptide collecting device

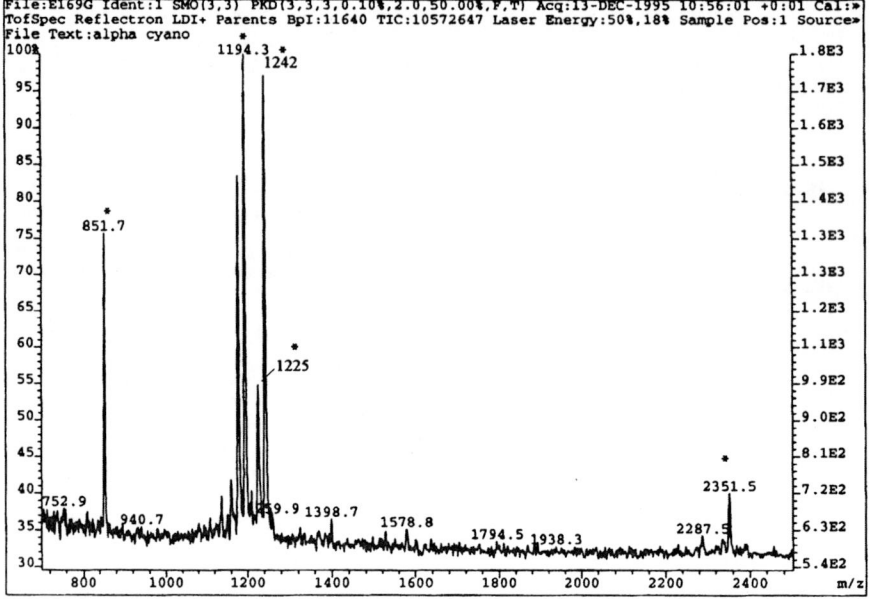

Fig. 5. MALDI-MS spectrum of a tryptic in-gel digest from human heart protein spots separated by HP-2-DE after concentration and desalting using the peptide collecting device. The protein was identified as myosin ligth chain 2 by five matches (marked with *asterisk*). The mass spectrum was recorded with a Fisons TofSpec instrument in the reflectron mode

gave a better signal to noise ratio because buffer salts are removed.

The preparation of peptide mixtures by in situ cleavages on membranes is simpler and not as time-consuming. Many proteins are simply immobilized on membranes, but the blotting procedure leads to considerable loss of yield during transfer from gel to membrane. Furthermore, with very hydrophobic proteins the yield of released peptides is low. In this case, enzymatic cleavage in the gel matrix is always preferable.

References

Aebersold R., Teplow D, Hood LE, Kent SBH (1986) Electroblotting onto activated glass membranes. J Biol Chem 261: 4229–4238

Aebersold RH, Leavitt J, Savedra RA, Hood LE, Kent SBH (1987) Internal amino acid sequence analysis of proteins separated by one- or two-dimensional gel electrophoresis after in situ protease digestion on nitrocellulose. Proc Natl Acad Sci USA 84:6970–6974

Eckerskorn C, Lottspeich F (1989) Internal amino acid sequence analysis of proteins separated by gel electrophoresis after tryptic digestion in polyacrylamide matrix. Chromatographia 28: 92–94

Fernandez J, DeMott M, Atherton D, Mische SM (1992) Internal protein sequence analysis: enzymatic digestion for less than 10 µg of protein bound to polyvinylidene difluoride or nitrocellulose membranes. Anal Biochem 201:255–264

Fernandez J, Andrews L, Mische SM (1994) An improved procedure for enzymatic digestion of polyvinylidene difluoride-bound proteins for internal sequence analysis. Anal Biochem 218:112–117

Hellman U, Wernstedt C, Gone`z J, Heldin CH (1995) Improvement of an "ingel"digestion procedure for the micropreparation of internal protein fragments for amino acid sequencing. Anal Biochem 224:451–455

Jenö P, Mini T, Moes S, Hintermann E, Horsz M, (1995) Internal sequences from proteins digested in polyacrylamide gels. Anal Biochem 224:75–82

Jungblut P, Otto A, Zeindl-Eberhart E, Pleissner KP, Knecht M, Regitz-Zagrosek V, Fleck E, Wittmann-Liebold B (1994) Protein composition of the human heart: the construction of a myocardial two-dimensional electrophoresis database. Electrophoresis 15:685–707

Kellermann J, Grimm R, Eckerskorn C (1995) "Enzymatische Spaltung von Proteinen im SDS-PAGE - teilweise Automatisierung der Probenvorbereitung". In: Kellner R, Lottspeich HE, Meyer HE (eds) 2. Arbeitstagung "Mikromethoden in der Proteinchemie", 26.-30. Juni 1995 Max-Planck Institut für Biochemie, Martinsried

Klose J (1975) Protein mapping by combined isoelectric focusing and electrophoresis in mouse tissues. A novel approach to testing for induced point mutations in mammals. Humangenetik 26: 231–243

Klose J, Kobalz U (1995) High resolution two-dimensional gel electrophoresis. Electrophoresis 16: 1034–1059

Matsudeira P (1987) Sequence from picomole quantities of proteins electroblotted onto polyvinylidene difluoride membranes. J Biol Chem 262:10035–10038

Otto A, Thiede B, Müller E-C, Scheler C, Jungblut P (1996) Identification of human myocardial proteins separated by 2-DE using an effective sample preparation for mass spectrometry. Electrophoresis (in press)

Pappin DJC, Hojrup P, Bleasby AJ (1993) Rapid Identification of proteins by peptide-mass fingerprinting. Current Biology 3:327–332

Rosenfeld J, Capdevielle J, Guillemot JC, Ferrara P (1992) In gel digestion of proteins for internal sequence analysis after one- or two-dimensional gel electrophoresis. Anal Biochem 203:173–179

Thiede B, Otto A, Zimny-Arndt U, Müller E-C, Jungblut P (1996) Identification of human myocardial proteins separated by two-dimensional electrophoresis with matrix-assisted laser desorption ionization mass spectrometry. Electrophoresis 17:588–599

Zeindl-Eberhart E, Jungblut P, Otto A, Rabes HM (1994) Characterization of tumor-associated protein variants during hepatocarcinogenesis in the rat. J Biol Chem 269:14589–14594

Isolation of Peptides for Microsequencing by In-Gel Proteolytic Digestion

U. Hellman

8.1
Background

In contrast to predictions made at the advent of molecular biology about 15 years ago, the demand for classical amino acid sequence analysis has increased. However, the goal no longer is to determine the complete primary structure of a protein by direct sequencing, but rather to sequence a few relatively short peptides derived by proteolytic digestion of the purified protein. The sequences obtained are usually utilized for the design of oligonucleotide probes for subsequent cDNA cloning and DNA sequencing. Another important use of internal peptide sequences is the design and production of anti-peptide antibodies which have proved to be an extremely useful analytical and preparative tool. Yet another application is the positive identification of purified proteins.

Today, when much of the analyses are made on minute amounts of sample, often obtained from cultured cells, the need for highly sensitive preparative techniques cannot be overemphasized. A general problem with commonly used column chromatography methods is the losses that frequently occur due to adsorption of proteins onto the column support and onto various surfaces such as pipette tips, tubing, and fraction tubes. This is typical for most samples and often seen with hydrophobic samples, e. g., membrane associated proteins.

A convenient preparative technique offering both high recovery micropurification as well as the possibility to handle practically any protein is to obtain the sample for proteolytic digestion as a Coomassie blue visible band in a polyacrylamide gel (Rosenfeld et al. 1992; Hellman et al. 1995). This procedure can be performed either early in a purification protocol or as a later step; the only demands are that: the sample must be present in amounts of at least 10–20 pmoles, be homogeneous, and definitely represent the protein in question and not a contaminant.

The following describes in detail the in-gel digestion procedure, as it has been used in our laboratory for well over 400 proteins.

8.2
Outline of the Procedure

1. Excise band(s); transfer to 1.5 ml Eppendorf tubes.

2. Wash out stain and SDS; replace destaining solution with digestion buffer.

3. Dry completely.

4. Allow protease to absorb into the polyacrylamide gel while rehydrating.

5. Incubate.

6. Extract peptides.

7. Isolate peptides by liquid chromatography.

8.3
Materials

Digestion with trypsin

– washing solution: 0.2 M ammonium bicarbonate, 50% acetonitrile

– incubation buffer: 0.2 M ammonium bicarbonate

– extraction solution: 0.1% trifluoroacetic acid, 60% acetonitrile

Digestion with LysC

– washing solution: 0.5 M Tris-HCl (pH 9.2), 50% acetonitrile

– incubation buffer: 0.1 M Tris-HCl (pH 9.2)

– extraction solution: 0.1% trifluoroacetic acid, 60% acetonitrile

Enzymes

– modified trypsin, sequence grade. (Promega Corp., Madison, WI, #V5111). We reconstitute 20 µg lyophilized batches in 200 µl of 1 mM HCl. The 0.1 µg/µl solution is aliquoted in 12 µl portions and stored at –20 °C. For each digestion 5 µl is added to the substrate. Since trypsin generates many and short fragments, it is mainly used for peptide mapping or if the aim of the analysis is to establish the identity of a protein.

– endoproteinase LysC (from *Achromobacter lyticus*) (WAKO Chemicals GmbH, Neuss, Germany, #129–02541). A stock solution, 5 mg/ml in 0.1 M Tris-HCl (pH 9.2), which is stored at –20 °C, is diluted 100-fold with the same buffer. The

0.05 µg/µl solution is aliquoted in portions of 25 µl and stored at –20 °C. For each digestion 10 µl is added to the substrate. We prefer to use the LysC specific protease for most digestions; it is a very specific and stable enzyme, generating relatively long fragments which facilitate the design of oligonucleotide probes. Due to its high specificity, we know that each fragment (except for the NH$_2$-terminal one) is preceded by a lysine residue, which in effect "prolongs" the peptide with a "two-codon" amino acid.

A good chromatograph, capable of performing micro- or narrow-bore reversed phase liquid chromatography, should be employed. We use the SMART system (Pharmacia Biotech, Uppsala, Sweden), which allows simultaneous monitoring of the eluate at three wavelengths, thereby providing important information about the peptides prior to sequence analysis. For primary isolation and rechromatography, respectively, we utilize the µRPC C2/C18 SC 2.1/10 and Sephasil C8 SC 2.1/10 columns. **Chromatographic equipment**

All isolations of proteolytic fragments obtained by in-gel digestion are carried out by reversed phase liquid chromatography. The peptides are eluted by a shallow linear gradient (0% to 40% in 160 min) of acetonitrile in 0.06% trifluoroacetic acid at a flow rate of 100 µl/min. Fractions are selected on basis of the differentiated UV signal and automatically collected into 0.5 ml Eppendorf tubes. Monitoring is done at 215, 254 and 280 nm. These wavelengths give valuable information regarding the nature of the eluting peptides, since the presence of Phe or pyridylethylcysteine (254 nm), Tyr or Trp (ratios 280/254/215) or contaminating aromatic compounds can easily be diguised. **Chromatographic conditions**

- trifluoroacetic acid, spectroscopic grade (Pierce, Rockford, IL #28902). **Solvents**

- acetonitrile, liquid chromatography solvent quality (Merck, Darmstadt, Germany #1.14291).

- water, Milli-Q quality. Do not allow water to come into contact with certain plastics such as PVC tubing which contains aromatic softeners.

- rotary evaporator (for drying the gel and reducing volumes). Alternatively, a slow stream of nitrogen can be used. **Miscellaneous**

- incubation device such as an air-heated column oven, equipped with a small shaker.

8.4
Procedure

1. Run the **reduced and alkylated** sample in a standard SDS-PAGE or 2-dimensional gel. Make sure that the protein is present in sufficient amount, i. e., > 10 pmole. Bands from several lanes may be combined, but if the sample is dilute, it is better to concentrate by an appropriate method and run fewer lanes. We prefer 4-vinylpyridine as alkylator, since it adds a chromophore to cysteine residues, making it possible to identify cysteine containing peptides by their specific UV absorption at 254 nm.

2. Stain and destain by standard Coomassie methods.

3. Cut out the band(s) positively known to be the appropriate ones. It is a good practice to run one empty lane for a negative control and one lane with a known amount of a standard protein (i.e., 4.5 µg of reduced and alkylated ovalbumin) for a positive control.

   Note: If the gel has been dried either plastic-wrapped or onto filter paper, it must first be rehydrated with water until it can be liberated from the support.

4. Wash the gel piece(s) twice with washing solution adding enough volume to cover the gels, usually 100–300 µl. Incubate with shaking for at least 30 min at 30 °C. Discard the washings. Washing time can for convenience be prolonged without any problem, as it is very difficult to wash out intact proteins.

5. Dry the gels **completely** under a stream of nitrogen or in a rotary evaporator. This is a small but important difference from the original procedure (Rosenfeld et al. 1992)

6. Start the rehydration/reswelling of the gels with a few microliters of digestion buffer.

7. While the gels are still swelling, add 0.5 µg of the protease so that it is absorbed **into** the gel pieces. Add the digestion buffer in small aliquotes so that the gel swells slowly to its original (or to an even larger) size.

8. Incubate at 30 °C overnight.

9. Acidify by adding trifluoroacetic acid (TFA) to a final concentration of 1%.

10. Collect the supernatant. Add extracting solution (enough volume to cover the gel) and incubate for at least 30 min at 30 °C. Combine extract with the previous supernatant and repeat extraction. Combine extracts. If preferred, incubations time may safely be prolonged.

11. Reduce volume by rotary evaporation to about one third; this removes most of the organic solvent from the extraction solution.

12. Isolate generated fragments by narrow-bore reversed phase liquid chromatography.

8.5
Results and Comments

The example (kindly provided by Dr. Lars Rönnstrand) in Fig. 1 shows a sample obtained by affinity chromatography using a column with an immobilized synthetic peptide. The desorbed material, about 20 µg in total, was subjected to SDS-PAGE and several interacting proteins of various molecular weights were seen as Coomassie stained bands. All were in situ digested with LysC, and the chromatogram resulting from a 200 kDa band is shown in the figure. After sequencing a few selected peptides, it was concluded that the 200 kDa protein is identical to myosin; an unexpected finding. Although about 95 peaks were seen on the chromatogram, it was possible to select several apparently homogeneous peptides, and the identity of the sample could be established unambigously without rechromatography. The automatic peak fractionation will, with properly set parameters, collect fractions conveniently. A representative example is given in Fig. 1B, where an enlarged portion (from 45 to 85 min) is shown with the determined sequences written out.

In spite of the powerful resolution obtained with modern dedicated chromatographic equipment, it sometimes happens that a fraction contains two or more peptides. This is usually revealed by the asymmetric shape of the peak or by the different maxima of the two extra UV signals, demonstrating heterogeneity in the aromatic content of the participating peptides. When this is suspected, we pool up to four poorly resolved fractions, reduce the concentration of organic solvent by rotary evaporation and rechromatograph the peptide mixture on the C8 column. The upper curve in Fig. 2 shows a portion of a chromatogram obtained after tryptic digestion of a histone preparation isolated by electrophoresis in a Triton acidic urea gel. The bot-

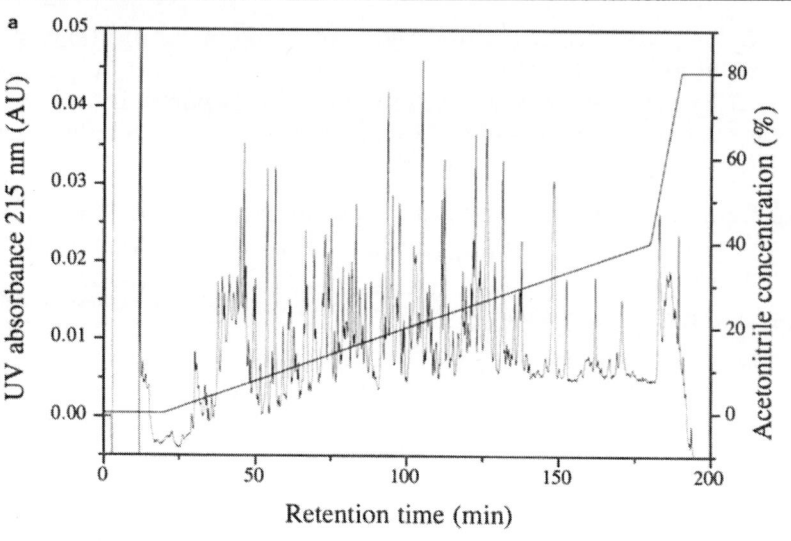

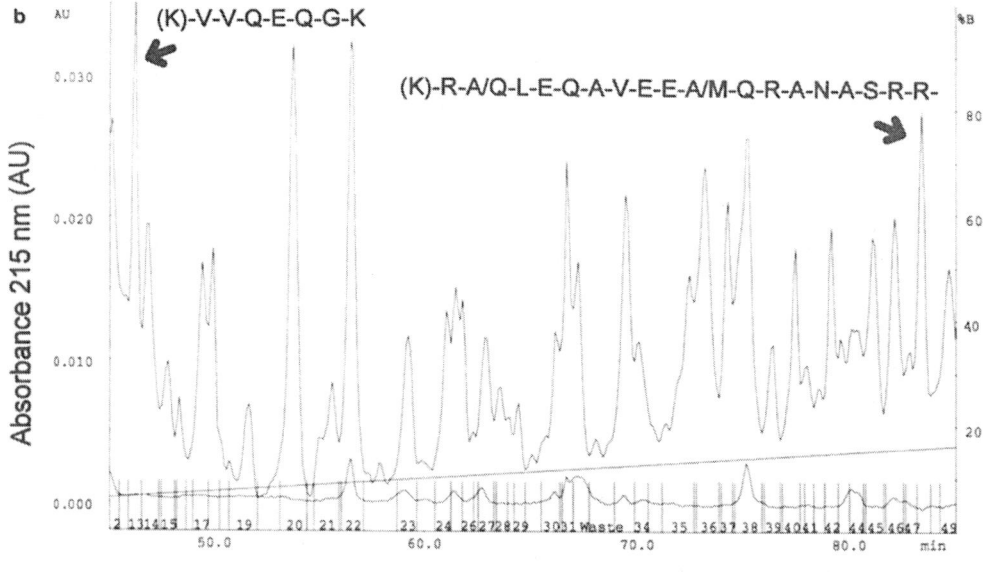

Fig. 1. a Isolation by reversed phase liquid chromatography of proteolytic fragments from a 200 kDa protein. The fragments were generated by in-gel digestion with a LysC specific protease. **b** Part of the previous chromatogram (45 to 85 min), showing fraction collection marks (*vertical bars*) as well as the amino acid sequences of two peptides. For further details, see text

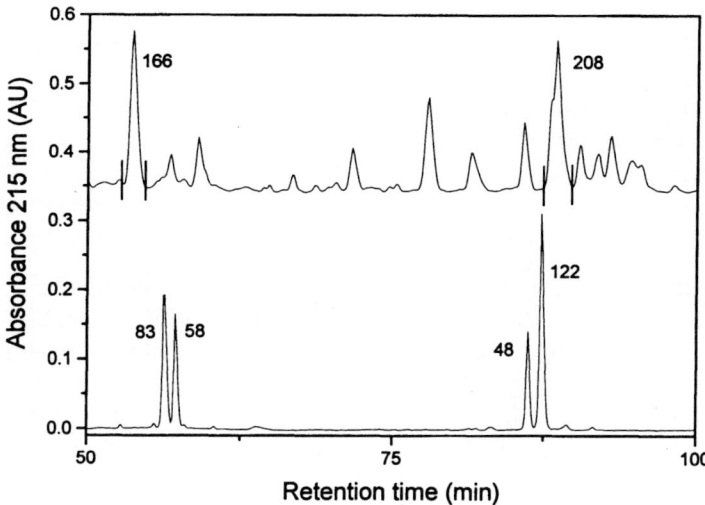

Fig. 2. Reversed phase liquid chromatography of a tryptic digest of a histone protein. For clarity only a selected part of the chromatogram is shown. The *upper curve* shows the digest fractionated on the C2/C18 column. In the *bottom curve*, the two marked fractions were pooled and rechromatographed on the C8 column. The figures given near the fractions are areas in arbitrary units as they are integrated by the SMART software

tom curve is the resulting separation obtained by rechromatography of a pool of the two marked fractions. The recoveries upon rechromatography of the two fractions are 85% and 82%, respectively. The peptides in all four resulting fractions were found to be homogeneous by sequence analysis.

Proteins in dried gels, either on filter papers or between sheets of plastic wrap, can also be in situ digested. This possibility offers analysis of proteins in archived, perhaps several year old gels, or from gels which for various reasons were dried, e.g., for autoradiography.

8.6
Troubleshooting

- If no peptides are obtained from a gel band with a sufficient amount of protein present, then the reason may be low activity of the enzyme. When this is suspected, repeat the whole procedure over again with a different enzyme preparation. In most cases positive results are obtained. This demonstrates that there is little risk of losing the intact sample even though the gel is subjected to several washing and extraction steps.

- It should be pointed out that when no peptides are obtained, the predominant reason is that the quantity of the sample is too low. It is unsafe to estimate amount of protein from the intensity of Coomassie staining since proteins stain with different efficiences. As pointed out earlier, it is advisable to include a known amount of a standard protein as a positive control, as well as to perform a blank digest in a gel piece without protein as a negative control.

- The recovery will be lower if the digestion is done in a large volume of gel. To overcome this, the sample must be concentrated prior to electrophoresis. The method of choise is dependent on the sample; good results have been obtained with the chloroform/methanol/water based precipitation protocol (Wessel and Flügge 1984).

- In our experience, if about 10–20 pmole of protein is **loaded** on a polyacrylamide gel, enough material for sequence analysis is generated. This is based on using an Applied Biosystems Model 470A sequencer; later models are specified with approximately fivefold higher sensitivity.

- If the recovery of generated peptides is high; redigestion of a successfully digested gel band does not result in additional peptides (Hellman et al. 1995).

The in-gel digestion procedure offers a rapid and nearly generally applicable technique to generate internal protein fragments from practically any protein.

References

Hellman U, Wernstedt C, Gonez J, Heldin C-H (1995) Improvement of an "in-gel" digestion procedure for the micropurification of internal protein fragments for amino acid sequencing. Anal Biochem 224: 451–455

Rosenfeld J, Capdeville J, Guillemot J C, Ferrara P (1992) In gel digestion of proteins for internal sequence analysis after one- or two-dimensional gel electrophoresis. Anal Biochem 203:173–179

Wessel D, Flügge U I (1984) A method for the quantitative recovery of protein in dilute solution in the presence of detergents and lipids. Anal Biochem 138:141–143

Part III
Manual and Automated Microsequencing Methods

Automated Microsequencing: Introduction and Overview

B. WITTMANN-LIEBOLD

9.1
Introduction

Frederick Sanger and coworkers were the first to determine the amino acid linear structure of a protein, namely of bovine insulin, with a total of 51 amino acids in two chains (Sanger and Thompson 1953) other insulins were explored shortly thereafter. Acid hydrolysis was used to generate short di-, tri and tetra-peptides which were isolated by paper chromatography and NH2-terminally labeled by 2,4-dinitrochlorobenzene (the Sanger reagent) with formation of the dinitrophenyl- (DNP-) peptide. After hydrolysis of the peptide and ether extraction of the labeled, yellow-colored, NH2-terminal DNP-amino acid, the first residue was identified by two-dimensional paper chromatography. By comparing all resulting peptides Sanger succeeded in the final alignment of the amino acids within the two polypeptide chains. For this work he was awarded the Nobel prize in 1958.

In 1950 Pehr Edman significantly advanced protein structure analysis by introducing a chemical technique for the sequential degradation of polypeptides from the NH2-terminal end. This degradation became famous as the Edman degradation (Edman 1956). With this method on hand several short peptide hormones were sequenced, such as melanotropin-α and -β (Harris and Lerner 1957; Harris and Roos 1956), ACTH (Li et al. 1955), the lysine and arginine vasopressins and oxytocin (by the laboratories of Du Vigneau and Acher). Finally, the amino acid analyzer was invented (Spackman et al. 1958), which for the first time allowed highly accurate, quantitative amino acid compositional analyses of polypeptides. It became possible to study the primary structures of large proteins and to complete their sequences, e.g., the α- and β-chains of adult human hemoglobin (Braunitzer et al. 1961) with 141 and 146 residues, respectively, cytochrome c (Margoliash et al. 1991), chicken lysozyme (Canfield 1963), bovine ribonuclease (Smyth et al. 1963), bovine chymotrypsinogen

A (Keil et al. 1963), tobacco mosaic virus coat protein (Wittmann-Liebold and Wittmann 1963) and sperm whale myoglobin (Edmundson 1965). Only proteins which were available in gram amounts and easy to purify could be completely sequenced. Limiting step were the tedious separation techniques of peptides by one-dimensional paper chromatography or paper electrophoresis and the lack of a sensitive amino acid analyzer and effective column chromatographic methods for the purification of proteins and peptides. In the beginning fraction collectors were not available and the fractions were collected manually. Then, cation exchange chromatography, using Dowex 50, was invented but this technique did not allow separation of basic or hydrophobic peptides. In fact only 50% of the tryptic hemoglobin peptides were recovered from the column under these conditions. An alternative was the use of an anion exchanger (Dowex 1) and elution of the peptides with volatile pyridine acetate buffers between pH 9.0 and 5.0 (Rudloff and Braunitzer 1961). However, the consequence was a strong pyridine smell in the laboratory!

In 1967 Edman and his colleague Geoffry Begg published the automation of the Edman degradation, facilitating extended sequencing of polypeptides to about 60 steps in a reasonable time with about 7 mg of purified protein. The machine, called a sequenator, allowed systematic investigation of protein and peptide structures and led to rapid increases in known protein sequences. Unfortunately Edman was never awarded the Nobel prize for this important invention, which was the key to modern protein structure analysis and which opened many new research fields in biology, biochemistry and modern medicine. At present more than 71 800 complete protein sequences are compiled in protein sequence databases (SwissProt and NBRF databases, release of January 1996) due to the combined approach of amino acid and nucleotide sequencing together with application of new mass spectrometric techniques. Further increases in the number of known proteins are to be expected.

Edman studied the degradation process (Fig. 1) very thoroughly and introduced phenylisothiocyanate (PITC) as a coupling reagent. The chemistry which he introduced has virtually remained unchanged up to now. The changes that have been made are mainly due to more sophisticated machine technology, allowing application of much smaller amounts of sample for the degradation. At present, only low picomole quantities (a few micrograms) are needed vs the milligram amounts used before, and extended sequences may be obtained using 50–100 picomoles of protein.

1. COUPLING

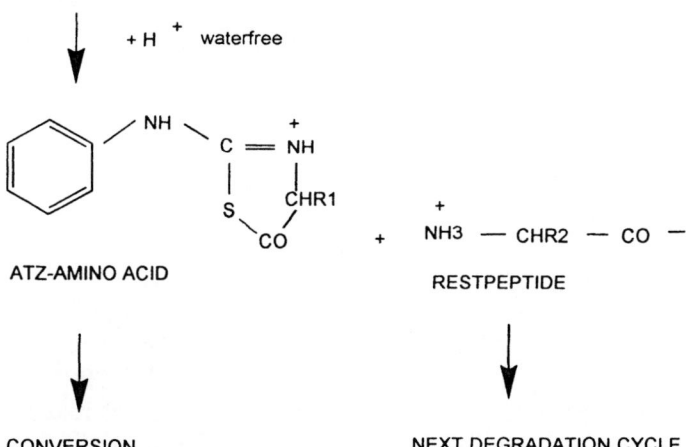

2. CYCLISATION

Fig. 1. (Continued on page 110)

3. CONVERSION

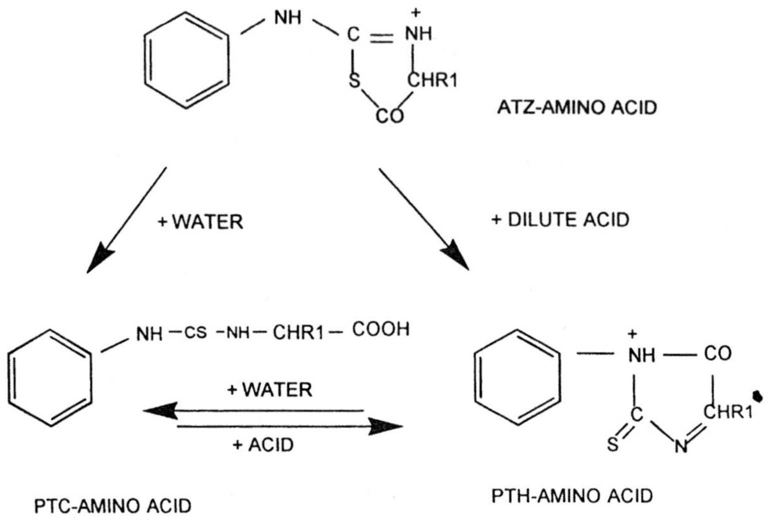

Fig. 1. Edman degradation scheme

9.2
Edman Degradation

9.2.1
Chemistry

The coupling reagent PITC reacts with the free α1 NH2-terminal group of the polypeptide chain under the formation of a phen-ylthiocarbamoyl peptide (PTC peptide). The coupling reaction is performed at 45°–48°C for about 15 min (vs 30–40 min in the early machines) and employs a large excess of reagent to drive the organic reaction to completeness. Treatment by a strong an-hydrous acid, e.g., trifluoroacetic acid (TFA), causes cyclization of the first residue thereby releasing this amino acid from the remaining polypeptide chain (cleavage) as a 2-anilino-thiazo-linone derivative (ATZ amino acid) (Fig. 1).

After the first round (cycle) of degradation, the remaining polypeptide chain is accessible to the next and successive cycles of degradation. Edman also introduced the conversion (iso-merization) of the ATZ amino acid derivative into the more sta-ble phenylthiohydantoin (PTH amino acid) by treatment with

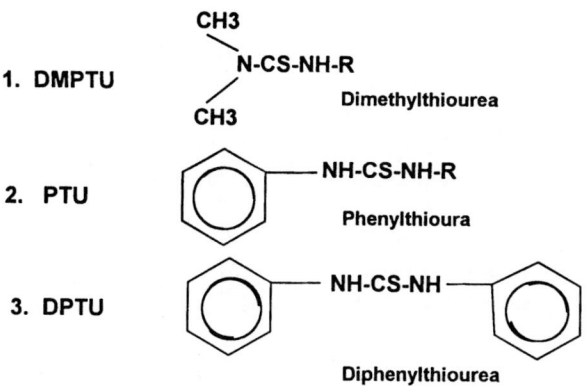

1. DMPTU

CH3
\
N-CS-NH-R
/
CH3   Dimethylthiourea

2. PTU

NH-CS-NH-R

Phenylthioura

3. DPTU

NH-CS-NH

Diphenylthiourea

Fig. 2. Side products of the Edman reaction. These byproducts are present in the HPLC-trace of the PTH-amino acids (see Fig. 9)

aqueous acid; but for many years this conversion had to be performed manually. Finally, it was automatized directing the released ATZ amino acid derivative into a conversion flask (Wittmann-Liebold et al. 1976). When the machines were adapted to smaller protein amounts the volumes of reagents and solvents had to be adjusted from milliliter amounts to microliter quantities and new valves as well as a smaller converter were introduced (Wittmann-Liebold et al. 1976; Wittmann-Liebold 1981).

Identification of the PTH amino acid derivatives was done by paper chromatography and staining with an iodine-starch reagent (Edman and Henschen 1975). Although the degradation in the Edman sequenator was effective, direct identification of the released PTH amino acids was more cumbersome due to the formation of side products of the reaction which obscured identification (see Fig. 2) by paper chromatography.

Even in modern microsequencing, the formation of byproducts cannot be avoided, due to water hydrolysis of the PITC reagent. For this reason, manual degradations were done in combination with amino acid rest-peptide analysis using an aliquot of the peptide after each step for acid hydrolysis and compositional analysis. Alternatively, the crude PTH amino acid was subjected to acid hydrolysis and the amino acid released analyzed in the amino acid analyzer (so-called back hydrolysis). At first, the Edman degradation was done by the paper strip method (Fraenkel-Conrat et al. 1955; Shelton and Schroeder 1960), in which peptide bands from preparative one-dimensional paper chromatography were subjected to the reaction with PITC in a dessicator with vapors of a volatile base (pyridine). Several paper strips of each peptide were treated with PITC, dried, and washed with benzene and ethyl acetate to remove the excess reagent. Cleavage was done in another dessicator with TFA vapors obtained by applying vacuum. The ATZ amino acid was extracted by ethyl ether and this extract discarded whilst one peptide strip per degradation cycle was eluted with 20% pyridine for amino acid compositional analysis in order to determine which amino acid was missing (subtractive Edman). Micromole amounts were necessary to perform the sequence analysis of a peptide at that time. Higher sensitivities were gained by applying the Edman degradation in combination with the dansyl technique: an aliquot of the degraded peptide was withdrawn for the reaction with dimethylaminonaphthalinesulfonyl chloride (dansylchloride) which yielded fluorescent derivatives (Gray and Hartley 1963). The resulting dansyl-peptide was subjected to acid hydrolysis, its NH2-terminal dansyl amino acid extracted and identified by two-dimensional thin-layer chromatography on polyamide sheets. The disadvantage was the loss of peptide sample at each of the cycles for this analysis. However, the fluorescent dansyl amino acid derivatives were detectable in the nanomole to subnanomole range. The introduction of the Chang reagent DABITC (dimethylaminoazobenzene isothiocyanate) (Chang and Creaser 1976) brought a new and innovative step towards higher sensitivity and quality of microsequencing by allowing direct identification of the released red-colored DABTH amino acids, visible by eye on 3×3 cm polyamide sheets in the nanomole to picomole range (see Chap. 10).

9.2.2
Edman-Based Sequencers

Microsequencing of proteins and peptides effectively is done in sequencers based on the Edman chemistry. Sequences are determined by a stepwise chemical procedure, degrading the amino acids of the polypeptide chain successively from the NH2-terminal end in the machine. Sequencers perform the degradation of higher repetitive yields (see below) and higher sensitivities can be gained than achieved in manual procedures. This is due to optimal exclusion of oxygen, good preservation of all chemicals in the machine under pressure of inert gas (argon or nitrogen), long durability and performance of the sequencer parts (Wittmann-Liebold, 1981, 1986, 1992) and the high sensitivity of UV detection (254 or 268 nm) of the released amino acid derivatives.

The Edman chemistry in the machines consists of three different reactions (Fig. 1):

- Coupling of the Edman reagent PITC to the α-amino group(s) of the protein under basic conditions

- Cleavage of the first residue from the remaining intact polypeptide chain as 2-anilinothiazolinone (ATZ amino acid) in strong anhydrous acid

- Conversion (isomerization) of the ATZ-amino acid in dilute acid to the more stable phenylthiohydantoin derivative (PTH amino acid) (Fig. 1b) within the converter

All these reactions are performed in a fully automatic mode and the cycles are repeated many times in order to read extended sequences. The released PTH amino acids are identified quantitatively *on-line* by high performance liquid chromatography (HPLC) (Wittmann-Liebold and Ashman 1985). Typically, between 10 and 100 picomoles of sample are applied onto the reactor of the sequencer. The amounts requested depend on the quality of the sample, the amino acid sequence, and whether short or extended sequences are desired. The quality of the degradation also largely depends on the purity of the sample, the

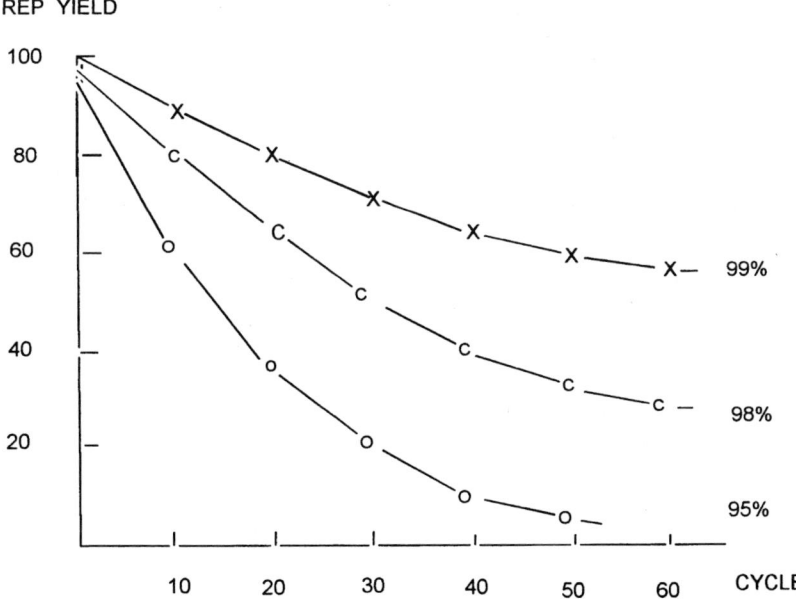

Fig. 3. Theoretically calculated repetitive degradation yields, given for a degradation of 95%, 98% and 99%. Sequencers typically reach 95%–98%; extended runs can only be interpreted if high repetitive yields of 98% are obtained. The theoretical yields may be considerably reduced if labile amino acids (Ser, Thr, Trp, etc.) occur or by wash-out of the sample

molecular mass, the individual sequence, and the technical constraints of the sequencer as well as on the quality of the chemicals. At best 50–60 residues can be identified unambigously. Often the sequence can be read further, but in the higher steps virtually all amino acids are present. Therefore, preceding cycles have to be substracted and reading the sequence requires extensive experience, and is likely, to be interpreted wrongly. Under optimal technical conditions the machine can reach a 96%–98% repetitive yield (Fig. 3) which means that at cycle 50 only a maximum yield of about 10%–20% of the amino acid residues in the 50th position can be expected together with up to 80%–90% of the previous PTH amino acids (the so-called overlap or lag) which obscures the identification. Hence, for extended runs the sequencer has to generate repetitive yields that are better than 96%. In order to judge the actual degradation and performance of the sequencer, the initial yield (that of the first released PTH amino acid) and the repetitive yield are calculated (Fig. 4).

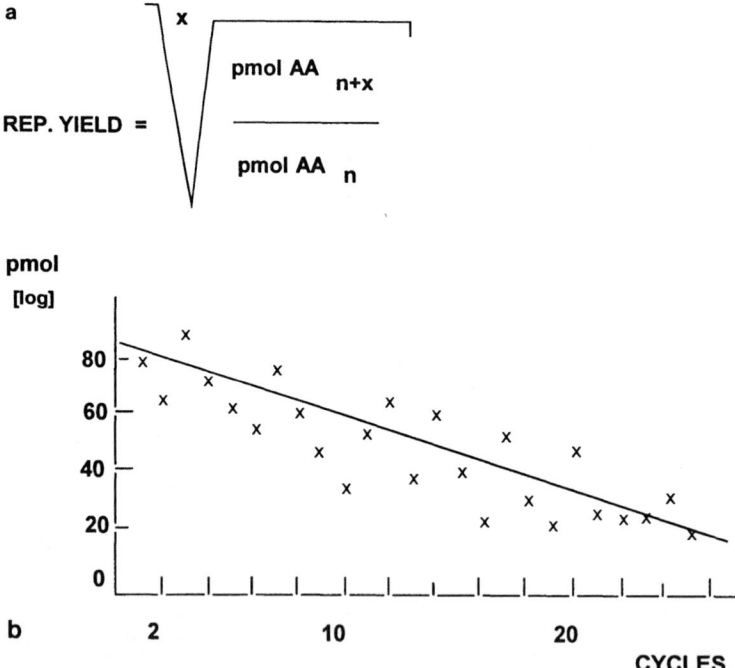

Fig. 4a,b. Calculation of the experimental repetitive yields of a run. Due to the different results obtained for each amino acid, only the yields of identical residues can be compared, e.g., the yield (in pmol) of Ala in step n with that of Ala in step n+x, etc., according to the formula given (a). Alternatively, the cycles vs the picomole quantities released in each cycle can be plotted (in logarithmic scale) and the extraplolation to step 1 indicates the repetitive yield (b)

The sequencers are equipped with software allowing quantitative determination of each PTH amino acid derivative in comparison to a standard PTH amino acid mixture. Furthermore, the software package enables corrections for background in the HPLC gradient chromatogram of the start cycle (without PTH amino acids) and/or substraction(s) of the previous cycle(s) that help to identify the next amino acid.

Most sequencers in use employ aqueous trimethylamine as the base for coupling of peptide with PITC, since this base is very volatile and can easily be removed by flushes of inert gas. In newer machines (e.g., the Procise sequencer) this base is replaced by N-methylpiperidine in water/methanol. After coupling with PITC the excess reagent and base must be removed completely by repeated washes with organic solvents, e.g., heptane and ethyl acetate, and extensive drying stages (flushs with inert gas). The cleavage is done in anhydrous TFA in order to avoid any hydrolytic cleavage of internal peptide bonds of the polypeptide chain. Such internal cleavages of peptide bonds would have a drastic influence on further degradation of the sample since new NH2-terminal groups might be formed which add to the background noise in the PTH amino acid chromatogram. Such hydrolytic cleavage of peptide bonds occurs in TFA containing water. Therefore, water in the TFA must be excluded completely. Refilling of a bottle under normal atmosphere already causes increases in water contamination of the TFA due to its hygroscopic properties. Large proteins have a particular tendency to cause severe background noise from newly released amino acids, since labile peptide bonds are more likely to occur (the weakest peptide bond is Asp-Pro).

After cleavage the acid has to be removed by extensive flushes of inert gas. The thiazolinone of the released amino acid then is dissolved in butylchloride and transferred into the converter where the conversion takes place in 25%–40% aqeuous TFA. After the conversion is done, the conversion medium is dried off and the PTH amino acid is dissolved in a suitable solvent mixture for automatic injection into the column of the HPLC system. Usually, sequencers are equipped with a gradient HPLC system but isocratic elution of the PTH-amino acids is also possible.

Important for high performance of the degradation is the use of high quality reagents and solvents (sequence grade chemicals), which must be freed of other components, mainly aldehydes, peroxides and radicals. Without this care the NH_2-terminal amino acids may be blocked during the cycles, and degradation may be stopped or the released amino acids destroyed. Further-

more, some sample washout occurs, mainly with short hydrophobic peptides. In these cases covalent attachment of the
polypeptides to suitable supports in combination with glass
membranes have been used (Laursen 1977; Wittmann-Liebold
1986; Machleidt et al. 1973). The support-bound peptide can be
filled into a column within which the Edman degradation takes
place (solid-phase sequencer, Laursen 1971) or the support-
bound peptide (e.g., a membrane-bound peptide) is introduced
into the cartridge, blot cartridge or flow-through reactor for the
degradation. The modification of the peptide with arylamine
membranes may be done directly in situ in the flow-through reactor (Herfurth et al. 1991a). Sample washout may also be reduced by treatment of the support with polybrene (Fig. 5) prior
to protein application. Polybrene is a quarternary base that helps

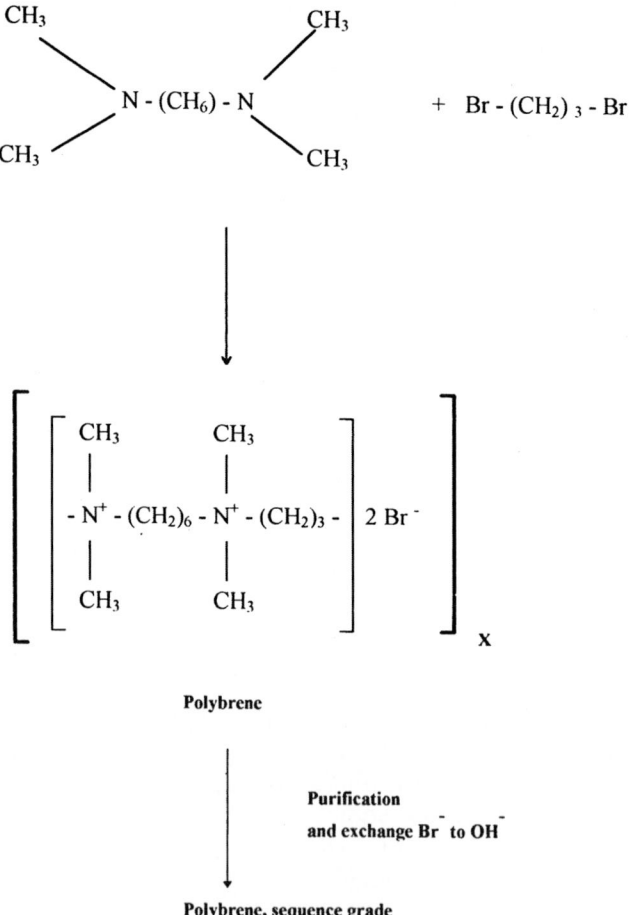

Fig. 5. Formation and structure of polybrene

the protein film stick to the glass or PVDF support. Previously, polybrene was introduced in order to fix the protein film to the glass wall of the spinning cup sequencer because the protein film was labile and could easily be cracked and/or washed out. However, polybrene does not form covalent bonds but provides electrostatic interactions with negatively charged residues of the proteins.

The construction of highly sensitive machines (Wittmann-Liebold 1981, 1986, 1992; Hewick et al. 1981) and their different installed modular parts will be discussed below. For the detection of the released amino acid derivatives several possibilities have been used in the past, among which were gas chromatography, various thin layer methods, and mass spectrometry. Finally, with the development of HPLC this technique became a routine for the PTH-amino acid identification (Zimmermann et al., 1977) and later, automated by adapting an *on-line* detection system (Wittmann-Liebold and Ashman, 1985) which then became commonly installed in all sequencers.

9.2.3
Materials

Sequencers consist of the following parts: **Equipment**

– reactor

– converter

– HPLC detection system

– valve systems

– heaters

– bottle system

– inert gas supply (either for argon or nitrogen)

– gas distribution system with meters, gauges, needle valves, etc.

– computer for storage and execution of the degradation

– conversion and HPLC programs.

The main part of the sequencer is the reactor in which the degradation takes place. Most commonly, the cartridge reactor version (Hewick and Hunkapiller 1981) is installed (Fig. 6a); alternatively, the blot reactor for blots, a column (e.g., for solid phase sequencing) or the flow-through reactor (Fischer et al. 1989) are used (Fig. 6b).

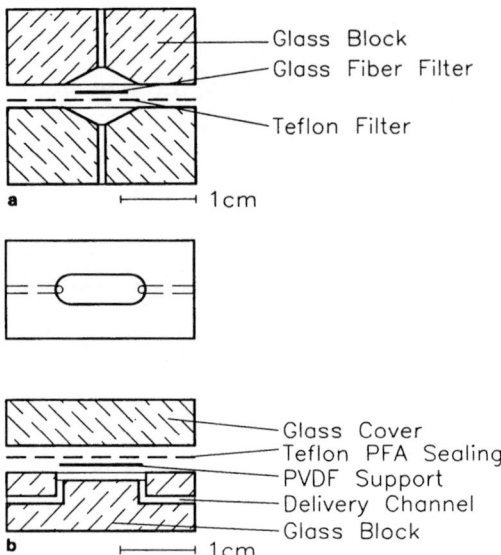

Fig. 6a,b. Sequencer reactors: a cartridge reactor (Hewick and Hunkapiller 1981); b flow-through reactor (Fischer et al. 1989) with front view (*top*) and cross-section (*bottom*)

The computer governs the uptake, storage and processing of the data. All the parts must be highly durable over long periods of time and they must be noncorrosive and nonreactive with the sequencing chemicals. Most critical is PITC and TFA, both of which slowly react even with the various perfluorated Teflon materials. Therefore, in some machines the essential parts within the valves are made of glass, e.g., in the Knauer machines. Furthermore, cross-contaminations of base and acid must be avoided. Hence, the gas supply of the bottles must be designed in a manner that avoids their interconnection, e.g., by the installation of different valves for pressurizing, venting and for delivery, one valve each per bottle (Fig. 7). This means that all bottles are equipped with three different valves which can be operated individually. Alternatively, by setting precise pressures with the appropriate electronic controls, the pressurizing valves may be omitted. In addition, waste and collect valves have to be installed to direct the exit fluids towards the waste bottle, or towards the converter or PTH analyzer system. The sequencer program includes operation of all these valves and controls the time settings. It directs the precise delivery of reagent and solvents and governs the programs for the main degradation executed within the reactor and for the side program in the converter. It also directs the start and execution of the HPLC program; maintains the tem-

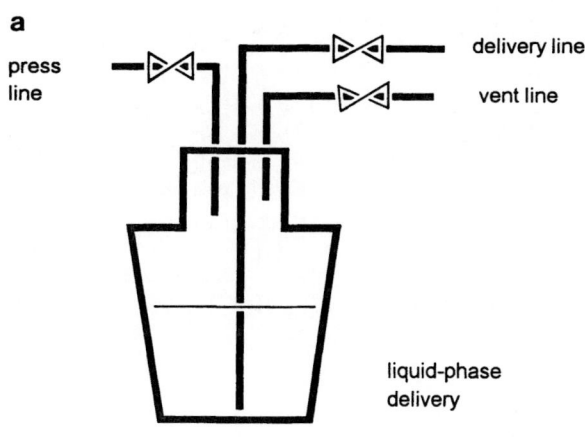

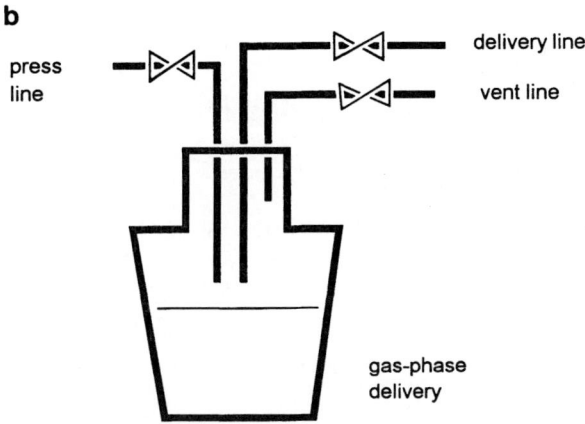

Fig. 7a,b. Liquid-phase delivery (a) vs gas-phase delivery (b) in the sequencer. Typically the base and/or the acid may be delivered in the gaseous mode (inlet and outlet line of inert gas above the reagent in the bottle) or in the liquid mode (outlet delivery line within the reagent solution, see text)

perature in the reactor, the converter and the HPLC column oven; and gives the signals for the different events, such as transfer of the fluid from reactor to converter, start of the converter program, transfer of PTH amino acid solution from the converter into the HPLC system, and the start of injection.

During the cycle repetitions all reagents and solvents are delivered in a fully automated manner. Moreover, they must be delivered precisely and in exactly the same amounts as in the previous cycles. In order to provide accurate deliveries, loops are designed

Reagents and solvents

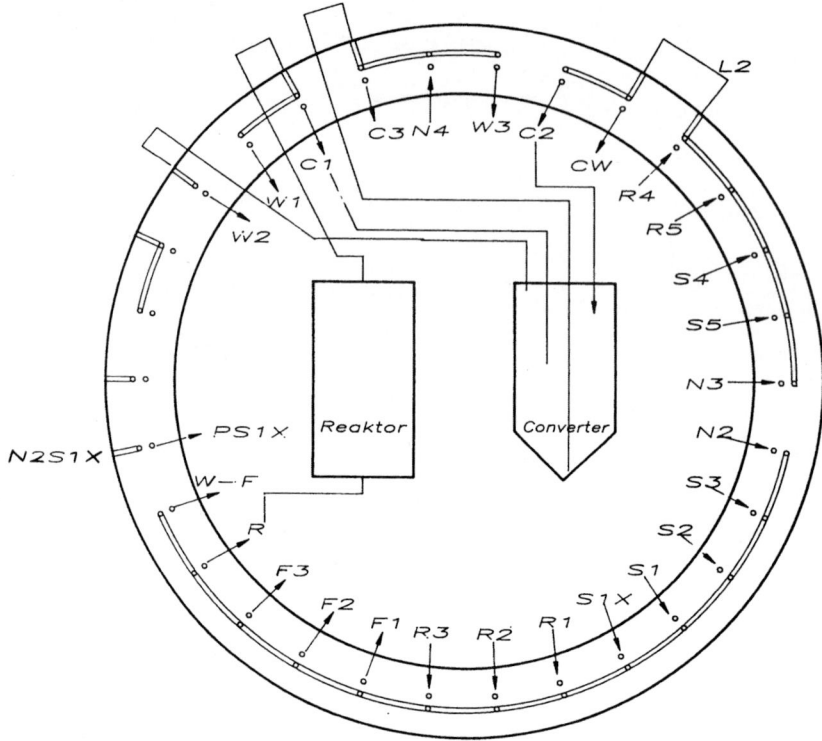

Fig. 8. Hybrid radiator valve system with central reactor and converter. All necessary valves (pressurizing-, ventilating-, delivery-, waste- and collecting valves) are arranged in one block with the reactor and converter in the center (arrangement of the valves in a new type of Berlin sequencer according to Horst Graffunder). Depending on the purpose, the radiator valve system may have a variable number of valve ports connected by central delivery line(s)

in combination with the valves (Wittmann-Liebold 1981). Alternatively, the inner geometry of the valve system (Wittmann-Liebold 1981; Fischer et al. 1989) is used to measure the volumes. Sequencers that are currently on the market, e.g., those from Knauer, Berlin and AppliedBiosystems/Perkin Elmer (Procise sequencer) utilize both approaches. The valves needed for pressurizing and venting the bottles, for deliveries into the reactor, converter and detection systems and for transfer of fluids are constructed in units (Wittmann-Liebold 1981) that have a common delivery line. These components can also be mounted within one block, e.g., in the hybrid radiator valve system (Fig. 8) (patent held by WITA GmbH), which minimizes the distances between the different valve units and the reactor and converter, which are centered within the circle.

The reagents (R) and solvents (S) which are incorporated in the sequencer are as follows:

- R1: 5% PITC in heptane
- R2: trimethylamine (TMA) in water (see below); alternatively N-methylpiperidine in water/methanol
- R3: anhydrous trifluoroacetic acid (TFA)
- R4: 25 or 40% TFA in water
- R5: standard mixture of PTH amino acids in acetonitrile, e.g., 25 pmol in 50 µl
- S1: n-heptane
- S2: ethyl acetate
- S3: 1-chlorobutane (butylchloride)
- S4: 20% acetonitrile in water; some suppliers add sodium acetate buffer
- S5: acetonitrile or methanol

As stated above all chemicals must be of high purity grade as measured by UV absorption, IR spectroscopy and rest-water content determination (Meinicke and Tschesche 1986); they can be purified by filtration through aluminium oxide (neutral) and redistilled from glass columns under nitrogen (see Meinicke and Tschesche 1986). Finally, the reagents and solvents must be tested in sequencer runs of standard proteins, e.g., β-lactoglobulin, ribonuclease, myoglobin or others, to guarantee the best sequencing performance.

Performed in two to three repetitions at 45°C and requiring about 20 min. It consists of the following:

Flow scheme

1. Delivery of the base, 12% TMA solution in water (R2), delivered in the gaseous mode (see above); alternatively, in the liquid mode, delivery of 4% TMA (R2) in water/methanol

Coupling

2. Delivery of 5% PITC in heptane (R1) in amounts that wet the sample filter completely, followed by short dryings

3. Repetition of the base delivery

4. Delay (reaction time) at 45°–48°C

5. Drying of the reagents by flushs of inert gas

6. Repeated washes with heptane (S1) and ethylacetate (S2), directed to waste

7. Drying

Thereafter the derivatized protein, the PTC peptide, is ready for cleavage.

Cleavage

1. Delivery of a few microliters of anhydrous TFA (R3)

2. 3 min of reaction at 45°–48°C (delay time)

3. Drying by inert gas flow to eliminate the acid

4. Delivery of butylchloride (S3) to dissolve the released phenylthiazolinone (ATZ amino acid)

5. Transport of the ATZ amino acid/butylchloride solution into the converter

6. Repetition of butylchloride extraction, collected into converter or discarded (directed to waste)

Conversion

1. Delivery of 25%–40% TFA in water (conversion medium, R4) into the converter

2. Conversion for about 10–20 min (including drying of the liquid) at 45°–50°C

3. Dissolution of the PTH amino acid in an appropriate solvent mixture (S4) (e.g., 20% acetonitrile in water with or without sodium acetate addition to adjust the pH), which allows direct injection into the HPLC column of the *on-line* HPLC system

4. Washing of the converter with solvent S4 or S5 (methanol or acetonitrile), directed to waste

On-line detection system

1. Transfer of the PTH amino acid solution via loop (for monitoring the volume) into the injection system

2. Transfer of an appropriate aliquot (50%–80%) of the total PTH solution into the column

3. Start of HPLC-gradient

4. Reequilibration of HPLC column to initial conditions prior to injection of the next amino acid

In the liquid mode the base is delivered as a liquid, whereas in the gaseous mode the reactor is only saturated with vapors of the base. In the liquid mode typical delivered amounts of base are 10–50 µl. In some machines the flow of the liquid is interrupted by pulses of inert gas (pulsed liquid mode), which avoids rapid rinsing of the base through the reactor thereby reducing washout of the sample. In the gaseous mode, inert gas (argon or nitrogen) pressurizes the flask, becomes saturated with base and water and transports this gaseous vapor mixture into the reactor. Thereby any washout of polypeptide sample, which has a tendency to dissolve in the base mixture, is circumvented. As a consequence, in the gaseous mode, the valves governing the delivery must be kept open during the entire reaction period whereas in the liquid mode, they are opened only briefly, during delivery of the liquid. Therefore, in the gaseous mode the 12% base is quickly depleted and must be replaced by a fresh bottle sooner or later, depending on the ambient temperature, in order to provide the appropriate conditions for the coupling reaction. In the liquid mode a few microliters of a 4% TMA solution (with the addition of 30% ethanol, methanol, ethylacetate or acetonitrile for wetting the PVDF filter) suffice to maintain a pH of 9.0 for the duration of the coupling reaction in the flow-through reactor (wet phase technique, Herfurth et al. 1991b). The method considerably reduces the amounts of R2 needed and can also be applied in the blot cartridge.

9.2.4
HPLC System

The *on-line* HPLC consists of two special pumps that allow low flow rates (about 200 µl) and which give rise to very little baseline noise or baseline shifts. The detector system must be capable of detecting the PTH amino acids in the low picomole range (microcuvettes). A typical HPLC separation of PTH amino acids (gradient mode) is given in Fig. 9.

Some sequencers employ analytical 4 mm (i.d.) C_{18} reversed-phase columns whereas others have 2 mm microbore columns. The solvent mixture (S4) is made of acetonitrile in water, with the addition of sodium acetate at nearly the same concentration and pH used in the initial separation conditions. The amount of organic solvent (acetonitrile) in this mixture is kept slightly below the start concentration (initial conditions) of the HPLC system in order to concentrate the injected PTH amino acid on top of the column. Therefore, the volume injected should not be too large, but must ensure dissolution of the PTH amino acid contained in the converter. Hence, volumes of 50–100 µl are necessary and about 50–75 µl are injected. For this reason, it is rather difficult to employ microbore columns which have inner diameters that are less than 2 mm. Usually only an aliquot of 50% of the total PTH amino acid solution is injected into the HPLC column in order to guarantee reproducible injections. Gas bubbles

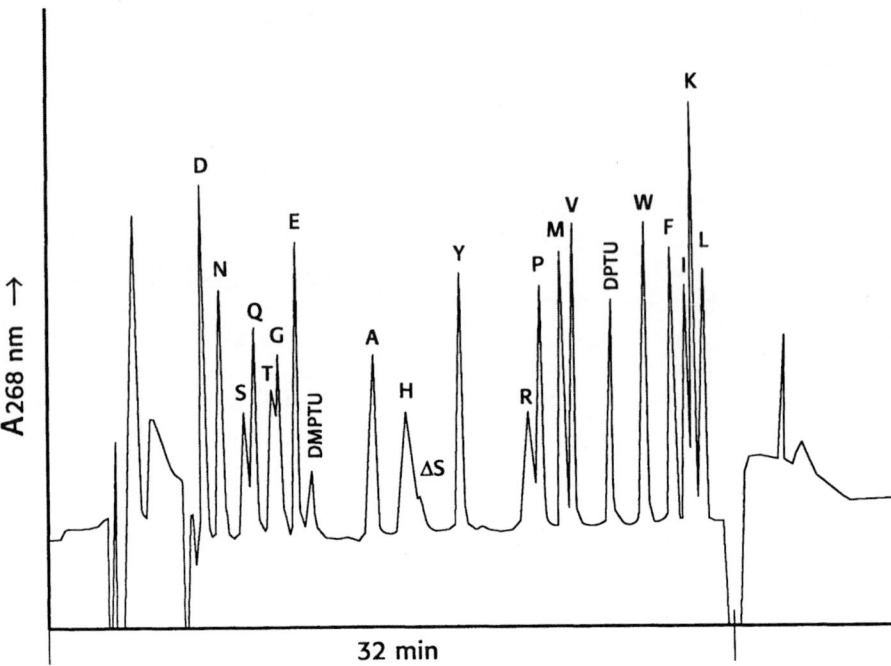

Fig. 9. Typical PTH amino acid standard chromatogram (sequencer Applied-Biosystems/Perkin Elmer, model 477A); 25 pmol per amino acid were injected. Gradient HPLC program as given in the manual of the machine

must be avoided during injection; therefore a delay time of a few seconds has to be programmed prior to the injection. It may happen that noninjection occurs; caused by long transfer distances between sequencer and analyzer, by spontaneously occurring bubbles, by blockage of the lines or by untightening of the inject valve or converter.

The HPLC gradient for the separation of the PTH amino acids is made from buffer A, e.g., 3% (Procise) or 5% (477A sequencer) tetrahydrofuran in water with the addition of 3 M sodium acetate buffer (pH 3.8 and pH 4.6, repectively) in a mixture that allows separation of histidine and arginine from all other PTH amino acids. Buffer B is acetonitrile (gradient grade); initial conditions vary from sequencer to sequencer, starting from 8% to 10% B. The separation is done in approximately 20–25 min plus reequilibration time of the column. Isocratic systems (Shimadzu and Berlin sequencer) employ a mixture of about 38% acetonitrile in water and sodium acetate. This solvent mixture can be recycled for about 2 months. Under these conditions the baseline is very stable and allows highly sensitive detections.

9.2.5
Degradation Programs

The entire process is governed by programs for:

- Preparation of the filter prior to sample loading (filter washes)

- The begin or start cycle, which carries out sample preparation and injection of the PTH amino acid standard mixture (begin cycles)

- The normal degradation cycles in the reactor (reactor cycle) and in the converter (conversion cycle), and execution of *on-line* detection by the HPLC run (PTH detection cycle)

Some sequencers have separate programs for both degradation in the reactor (e.g., in the cartridge, blot chamber or flow-through reactor) and in the converter, for the uptake of the released PTH amino acid, its transport from the converter into the column of the detection system and for the HPLC separation (detection program). Therefore, at stages in which collection of the thiazolinone in the reactor and transport into the converter take place, both the reactor and converter programs must be linked together. The same applies for transport of the PTH amino acid from the converter into the *on-line* HPLC system. As a result the times necessary for the individual programs have to be adapted to each other so that the collection event always occurs at the same stage within the program in all cycles performed. At the beginning of each sequencer run, the machine calculates the times required for each of the simultaneously running programs and inserts a delay time (by performing an argon or nitrogen flush) in order to start the conversion and *on-line* detection programs at the correct stage. This type of programming is found in the Perkin Elmer (Applied Biosystems) sequencers; however, the Berlin and Knauer sequencers have software that includes governing of the degradation, conversion and *on-line* detection events simultanously in one program so that no adaptation of the final times are necessary.

The precycles (filter washes) are designed to prepare the support (glass fiber filter or PVDF membrane) with polybrene (see above). After addition of this compound, the filter is dried and washed with a cycle that includes delivery of all reagents and solvents such that all active groups on the filter are saturated by the chemicals. This helps to decrease the background in the final chromatograms. Usually two to three filter washes are performed

Filter washes

prior to sample addition. This means that the reactor has to be reopened after the precycles for sample application.

Begin cycles   The begin cycle typically starts with delivery of TFA, drying by gas flushes and a solvent wash of the sample within the reactor. At the same time a standardized PTH amino acid mixture is applied into the converter, dried, dissolved in an appropriate solvent mixture suitable for injection into the HPLC column and then injected into the HPLC detection system in order to provide a typical standard HPLC chromatogram prior to the degradation cycles. Often a real blank conversion is performed with the PTH amino acid mixture in order to assess destruction of the amino acid derivatives. The standard chromatogram is needed in order to monitor the machine and to calibrate the PTH amino acid amounts. Aliquots of 25–50 pmol per PTH amino acid are delivered into the converter and 50%–80% is injected into the column. Quantification is done routinely by the software, which also handles the baseline corrections and subtractions of preceding cycles (rough data vs refined data).

Normal degradation programs   The normal degradation cycles of the reactor (Table 1), the converter (Table 2), and the *on-line* detection consist of the coupling, washings, cleavage, transport of the thiazolinone into the converter, conversion, dissolution of the PTH amino acid, transport into the HPLC detection system and the appropriate washes, reaction time delays and drying steps. In all of these steps oxygen must be completely excluded, and care must be taken not to contaminate reagents and solvents either with each other (cross-contaminations), with any outside contaminants or with oxygen from the atmosphere. Accordingly, all reagents and solvents must be of highest purity (sequencer grade), filled into the bottles under inert gas; the bottles must be kept under inert gas at all times.

Future design of sequencers   The need for sequencers which could perform at the highest level of sensitivity resulted in the development of machines that were suitable for microsequencing in the low picomole range. Due to this high sensitivity, internal sequencing of protein fragments from gels and blots has become routine in many laboratories worldwide. Finally, further developments in the microsequence analysis of polypeptides will lead to new techniques allowing amino acid sequencing in the femtomole to attomole range. In addition, several other strategies to determine the primary structure of proteins and polypeptides have become available (see below).

Table 1. Normal degradation progam in the cartridge (Procise sequencer)

Cartridge cycle: cart pulsed-liquid 1

Step	Function	Function name	Time/temp	Elution time
1	258	Begin	0	:00
2	137	Flush Input Block	5	:05
3	11	Del R2g, Cart (top)	20	:25
4	140	Flush Large Loop (Cart)	10	:35
5	6	Load R1, Cart (lg loop)	20	:55
6	131	Dry Cart (top)	30	1:25
7	140	Flush Large Loop (Cart)	5	1:30
8	135	Flush Cart Reagent Block	5	1:35
9	11	Del R2g, Cart (top)	170	4:25
10	140	Flush Large Loop (Cart)	5	4:30
11	6	Load R1, Cart (lg loop)	20	4:50
12	131	Dry Cart (top)	30	5:20
13	140	Flush Large Loop (Cart)	5	5:25
14	135	Flush Cart Reagent Block	5	5:30
15	11	Del R2g, Cart (top)	170	8:20
16	140	Flush Large Loop (Cart)	5	8:25
17	6	Load R1, Cart (lg loop)	20	8:45
18	131	Dry Cart (top)	30	9:15
19	140	Flush Large Loop (Cart)	5	9:20
20	135	Flush Large Reagent Block	5	9:25
21	11	Del R2g, Cart (top)	170	12:15
22	146	Wash Large Loop (Cart)	10	12:25

Table 1 (continued)

Cartridge cycle: cart pulsed-liquid 1

Step	Function	Function name	Time/temp	Elution time
23	140	Flush Large Loop (Cart)	10	12:35
24	135	Flush Cart Reagent Block	10	12:45
25	131	Dry Cart (top)	60	13:45
26	142	Set Cart Temperature	48	13:45
27	53	Del S2, Cart (sensor)	25	14:10
28	257	Wait	5	14:15
29	51	Del S2, Cart (top)	5	14:20
30	257	Wait	5	14:25
31	51	Del S2, Cart (top)	5	14:30
32	257	Wait	5	14:35
33	51	Del S2, Cart (top)	5	14:40
34	257	Wait	5	14:45
35	131	Dry Cart (top)	30	15:15
36	63	Del S3, Cart (sensor)	15	15:30
37	257	Wait	5	15:35
38	61	Del S3, Cart (top)	5	15:40
39	257	Wait	5	15:45
40	61	Del S3, Cart (top)	5	15:50
41	257	Wait	5	15:55
42	61	Del S3, Cart (top)	5	16:00
43	257	Wait	5	16:05
44	131	Dry Cart (top)	60	17:05

Step	Code	Description	Value	Time
45	111	Wash Input Block (S3)	10	17:15
46	137	Flush Input Block	30	17:45
47	140	Flush Large Loop (Cart)	5	17:50
48	26	Load R3, Cart (lg loop)	50	18:40
49	30	Transfer R3, Cart (gas)	5	18:45
50	136	Flush Cart Solvent Block	10	18:55
51	144	Wash Cart Solvent Block	10	19:05
52	136	Flush Cart Solvent Block	10	19:15
53	143	Wash Cart Reagent Block	15	19:30
54	135	Flush Cart Reagent Block	30	20:00
55	111	Wash Input Block (S3)	10	20:10
56	137	Flush Input Block	30	20:40
57	107	Wash Output Block (S2)	10	20:50
58	138	Flush Output Block	30	21:20
59	140	Flush Large Loop (Cart)	10	21:30
60	146	Wash Large Loop (Cart)	10	21:40
61	140	Flush Large Loop (Cart)	30	22:10
62	136	Flush Cart Solvent Block	60	23:10
63	257	Wait	95	24:45
64	131	Dry Cart (top)	40	25:25
65	142	Set Cart Temperature	48	25:25
66	127	Ready Transfer to Flask	0	25:25
67	141	Flush Transfer Line	5	25:30
68	63	Del S3, Cart (sensor)	15	25:45
69	257	Wait	20	26:05

Table 1 (continued)

Cartridge cycle: cart pulsed-liquid 1

Step	Function	Function name	Time/temp	Elution time
70	121	Transfer to Flask (gas)	30	26:35
71	141	Flush Transfer Line	5	26:40
72	53	Del S2, Cart (sensor)	25	27:05
73	257	Wait	20	27:25
74	121	Transfer to Flask (gas)	30	27:55
75	128	Transfer Complete	0	27:55
76	53	Del S2, Cart (sensor)	25	28:20
77	257	Wait	5	28:25
78	131	Dry Cart (top)	75	29:40
79	259	End	0	29:40

Table 2. Normal degradation progam in the converter (Procise sequencer)

Flask cycle: Flask normal 1

Step	Function	Function name	Time/temp	Elution time
1	258	Begin	0	:00
2	235	Set as Residue Cycle	0	:00
3	218	Flush Large Loop (Flask)	10	:10
4	173	Load S4, Flask (lg loop)	15	:25
5	213	Dry Flask	10	:35
6	218	Flush Large Loop (Flask)	10	:45
7	228	Ready to Receive	1	:46
8	213	Dry Flask	150	3:16
9	236	Pre-Conversion Dry	80	4:36
10	218	Flush Large Loop (Flask)	10	4:46
11	153	Load R4, Flask (lg loop)	20	5:06
12	213	Dry Flask	10	5:16
13	218	Flush Large Loop (Flask)	10	5:26
14	173	Load S4, Flask (lg loop)	15	5:41
15	218	Flush Large Loop (Flask)	10	5:51
16	257	Wait	540	14:51
17	226	Load Position	1	14:52
18	227	Prepare Pump	1	14:53
19	237	Post-Conversion Dry	200	18:13
20	213	Dry Flask	360	24:13
21	173	Load S4, Flask (lg loop)	15	24:28
22	213	Dry Flask	10	24:38

Table 2 (continued)

Flask cycle: Flask normal 1

Step	Function	Function name	Time/temp	Elution time
23	218	Flush Larg Loop (Flask)	10	24:48
24	173	Load S4, Flask (lg loop)	15	25:03
25	213	Dry Flask	10	25:13
26	218	Flush Large Loop (Flask)	10	25:23
27	221	Flush Injector	80	26:43
28	257	Wait	10	26:53
29	238	Concentrate Sample	5	26:58
30	225	Load Injector	40	27:38
31	171	Del S4, Flask	15	27:53
32	213	Dry Flask	10	28:03
33	212	Bubble Flask	5	28:08
34	215	Empty Flask	20	28:28
35	171	Del S4, Flask	10	28:38
36	213	Dry Flask	10	28:48
37	212	Bubble Flask	5	28:53
38	222	Flush Flask/Injector	40	29:33
39	221	Flush Injector	40	30:13
40	259	End	0	30:13

Partial amino acid sequencing leads to valuable sequence information which can be used for localization and sequence determination of the protein's gene, thereby facilitating final deduction of the amino acid sequence from the gene. However, it should be pointed out that this information is not sufficient to fully describe the primary structure of a protein! Whether the sequence really starts and ends at the deduced sequence, e.g., with or without expression of NH_2-terminal methionine, and whether the reading frame is correctly assigned may not be certain. Furthermore, the sites of modifications, e.g., glycosylations or phosphorylations, cannot be deduced from the nucleotide sequence. In order to determine such modifications and their functional implications, the protein has to be isolated, purified, and characterized by direct protein studies. The combined approaches to gain sequence information are as follows:

Comments

- partial NH_2-terminal or internal amino acid sequencing

- selection of suitable oligonucleotide probes

Genetic approach

- hybridization with genomic DNA

- localization of the gene for the protein

- cloning and sequencing of the gene

- deduction of the complete amino acid sequence from the gene sequence

- check of the correct open reading frame

- NH_2-terminal sequence analysis;

- protein cleavages to generate peptide fragments for internal sequencing

Sequence analysis by direct amino acid sequencing

- purification of peptides

- NH_2-terminal sequence analysis of the peptides

- additional fragmentation employing other enzymes

- separation and sequence analysis of these sets of peptides

- sequence alignment

- determination of S-S-bridges

- check for modifications

- Electrospray mass spectrometry (ES-MS)

Sequence analysis by mass spectrometry

- matrix-assisted laser-desorption mass spectrometry (MALDI-MS)

Mass spectrometry of proteins and peptides has become increasingly important since the development of ES-MS and MALDI-MS. Both techniques are applicable to the mass determination of proteins and peptides (see Chap. 19). Recently, these techniques, in combination with NH_2-terminal and internal amino acid sequencing, have allowed identification of protein variants and tumor-associated proteins from malignant tissues or biopsies after high resulution two-dimensional gel electrophoresis, e.g., of myocardium (Jungblut at al. 1994). These new methods encourage advancements in the field of protein research. By focusing directly on cellular proteins, they serve as an important complement to DNA studies.

Due to the increased amount of sequence data and the significant technical improvements in protein analysis (high resolution 2DE, HPLC chromatography, NH_2-terminal and internal sequencing by the Edman technique and mass spectrometry), more detailed studies in developmental biology and molecular medicine have become possible. Subjects of current interest include:

- tissue comparison
- identification of clinical markers
- multiple gene changes
- assessment of the role of posttranslational modifications in disease

References

Braunitzer G, Gehring-Müller R, Hilschmann N, Hilse K, Hobom G, Rudloff V, Wittmann-Liebold B (1961) The structure of adult normal human hemoglobin. Biol Chem Hoppe-Seyler 325: 283–286

Canfield R (1963) Peptides derived from tryptic digestion of egg white lysozme. J Biol Chem 238: 2698–2707

Chang JY, Creaser EH (1976) A novel manual method for protein sequence analysis. Biochem J 157: 77–85.

Edman P (1950) Acta Chem Scand 4: 283

Edman P, Begg G (1967) A protein sequenator. Eur J Biochem 1: 80

Edman P, Henschen A (1975) Sequence determination. In: SB Needleman (ed) Protein sequence determination. Springer, Berlin, Heidelberg, New York, pp. 232–279

Edmundson AB (1965) Amino acid sequence of sperm whale myoglobin. Nature 205: 883–887

Fischer S, Reimann F, Wittmann-Liebold B (1989) A new modular sequencer. In: B. Wittmann-Liebold (ed), Methods in protein sequence analysis. Springer, Berlin, Heidelberg, New York, pp. 98–107

Fraenkel-Conrat H., Harris JI, Levy AL (1955) in: Methods in biochemical analyis, vol. 2, p, 393

Gray WR, Hartley BS (1963) A fluorescent endgroup reagent for proteins and peptides. Biochem J 89: 59

Harris JI, Lerner AB (1957) Amino acid sequence of the α-melanocyte-stimulating hormone. Nature 179: 1346–1347

Harris JI, Roos P (1956) Amino acid sequence of a melanophore-stimulating peptide. Nature 178: 90

Herfurth E, Hirano H, Wittmann-Liebold B (1991a) The amino acid sequence of the *Bacillus stearothermophilus* ribosomal proteins S17 and S21 and their comparison to homologous proteins of other ribosomes. Biol Chem. Hoppe-Seyler 372: 955–961

Herfurth E., Pilling U, Wittmann-Liebold B (1991b) Microsequencing of proteins and peptides in the Knauer sequencer with and without covalent attachment to polyvinylidene difluoride membranes by the wet-phase degradation technique. Biol Chem Hoppe-Seyler 372:351–361

Hewick RM, Hunkapiller MW, Hood LE, Dreyer WJ (1981) A gas liquid solid phase peptide and protein sequencer. J Biol Chem 256: 7990–7997

Jungblut P, Otto A, Zeindl-Eberhart E, Pleißner KP, Regitz-Zagrosek V, Fleck E, Wittmann-Liebold B (1994) Protein composition of the human heart: the construction of a myocardial two-dimensional electrophoresis database. Electrophoresis 15:685–707

Keil B, Prusik Z, Sorm F (1963) Disulphide bridges and a suggested structure of chymotrypsinogen. Biochim Biophys Acta 78: 559–561.

Laursen RA (1971) General description of original sequencer and preparation of polystyrene. Eur J Biochem 20: 89–102.

Laursen RA (1977) Coupling techniques in solid-phase sequencing. Meth Enzymol 47: 277–288

Li CH, Geschwind II, Cole RD, Raacke ID, Harris JI , Dixon JS (1955) Amino acid sequence of alpha-corticotropin. Nature 176: 687–689

Machleidt W, Wachter E, Scheulen M, Otto J (1973) Coupling proteins to aminopropyl glass. FEBS Lett. 37: 217–220

Margoliash E, Smith EL, Kreil G, Tuppy H (1961) Amino acd sequence of horse heart cytochrome C. Nature 192: 1121–1127

Rudloff V, Braunitzer G (1959) Ein Verfahren zur präparativen Gewinnung natürlicher Peptide, die Isolierung der tryptischen Spaltprodukte von Humanhämoglobin A über Dowex 1X2 unter Verwendung flüchtiger Puffer. Biol Chem Hoppe-Seyler 323: 129–144

Pappin DJC, Hojrup P, Bleasby AJ (1993) Rapid identification of proteins by peptide mass fingerprinting. Curr Biol 3:327–332

Sanger, F, Thompson EOP (1953) The amino acid sequence in the glycyl chain of insulin. Biochem J 53: 353–374

Shelton JR, Schroeder WA (1960) Further NH2-terminal sequences in human hemoglobin A, S and F by Edman's phenylisothiohydantoin method. J Am Chem Soc 82: 3342

Smyth DG, Stein WH, Moore S (1963) The sequence of amino acid residues in bovine pancreatic ribonuclease: revisions and confirmations. J Biol Chem 238: 227–234

Spackman DH, Stein WH, Moore S (1958) Anal Chem 30: 1190.

Wittmann-Liebold B (1992) High sensitive protein sequence analysis. Pure and Appl Chem 64: 537–543

Wittmann-Liebold B (1986a) Design of a multipurpose sequencer. In: JE Shively (ed), Methods in protein microcharacterization. Humana, Clifton, NJ, pp 249–277

Wittmann-Liebold B (1986b) Practical methods for solid-phase sequence analysis. In: A. Darbre (ed) Practical protein chemistry: a handbook. Wiley, pp 375–409

Wittmann-Liebold B (1981) Microsequencing by manual and automated methods as applied to ribosomal proteins. In: TY Liu, AN Schechter, RL Heinrikson, PG Condliff (eds) Chemical synthesis and sequencing of peptides and proteins. Elsevier, Amsterdam, pp 75–110.

Wittmann-Liebold B, Ashman K (1985) The use of on-line high performance liquid chromatography for phenylthiohydantoin amino acid identification: In: H. Tschesche, (ed) Methods of protein chemistry. Walter deGruyter, Berlin, p 303

Wittmann-Liebold B, Graffunder H, Kohls H (1976) A device coupled to a modified sequenator for the automated conversion of anilinothiazolinones into PTH-amino acids. Anal Biochem 75: 621–633

Wittmann-Liebold B, Wittmann HG (1963) Die primäre Proteinstruktur von Stämmen des TMV: Aminosäuresequenzen des Proteins des TMV-Stammes Dahlemense. Z. Vererbungslehre 94: 427–435

Zimmermann CL., Appella A, Pisano JJ (1977) Rapid analysis of amino acid phenylthiohydantoins by high performance liquid chromatography. Anal Biochem 77: 569–573

Chapter 10

A Manual Method for Protein Sequence Analysis Using the DABITC/PITC Method

T. CHOLI-PAPADOPOULOU, Y. SKENDROS, and K. KATSANI

10.1
Introduction

Microsequencing of polypeptides by the Edman degradation technique can be performed either manually or by automated methods (Edman and Begg 1967; Edman and Henschen 1975). The degradation consists of three parts, coupling with 4-N, N'-dimethylaminoazobenzene-4'-isothiocyanate/phenylisothiocyanate (DABITC/PITC), acid cleavage of the first amino acid as phenylthiazolinone from the peptide chain, and isomerization to the more stable phenylthiohydantoins. The stable derivatives are identified on thin-layer polyamide sheets (Edman and Henschen 1975) or by high performance liquid chromatography (HPLC) (Chang 1988).

Like many other analytical procedures, manual sequencing has increased in speed and sensitivity over the years. Improvements in sensitivity are due to instrumentation, particularly HPLC, which has caused a small revolution in peptide mapping and purification as well as in analysis of amino acid derivatives. Some of this improvement is also a result of chemistry – choice of catalyst and small volumes in small reaction vessels that reduce drying times, simplification of handling procedures and assembly of all necessary components into a work station. Sensitive manual methods may also employ PITC, but in this case an HPLC detection system for identification of the released phenylthiohydantoins is required. Usually the derivatives are contaminated by UV-positive side products of the reaction that obscure the identification by thin-layer chromatography.

Alternatively, manual methods make use of isothiocyanate homologs that allow visual detection of the released amino acids. DABITC has been widely applied to manual liquid and solid phase microsequencing (Wittmann-Liebold and Kimura 1984). The reagent was described by Chang et al. in 1976 and has several advantages over the usual degradation with PITC: Firstly, it releases the amino acid thiohydantoins as red-colored derivatives which are visible to the eye in picomole quantities on polyamide sheets. Secondly, the side products of the Edman chemistry form blue-colored components easily differentiated from the intense red amino acid hydantoins. The true DABTH derivatives are red after HCl development, as described above, but other derivatives forming red, blue and purple spots also occur.

A disadvantage of DABITC, however, is that, due to the bulky side group of the DABITC reagent, the coupling yields are lower than those of PITC (Chang et al. 1978). Since introduction of the DABITC/PITC double coupling procedure, the method has found wide application:

- It is simple to perform, requires little sophisticated equipment and can be employed in any laboratory.

- Batch-wise processing, which is especially helpful in screening or characterizing HPLC fractions, can be done.

- It allows efficient sequencing of small peptides.

- It provides great flexibility in handling peptides, including both the occasional use of special procedures and the incorporation of improvements in methodology.

- The material is easy to handle and the method can easily be used for purity tests of isolated proteins or fragments.

The procedure can be applied to 1–5 nanomol peptide or protein, and with slight modifications is also suitable for picomole quantities. Even for minute amounts of peptide material the use of a sophisticated HPLC system is not mandatory.

Identification can most quickly and simply be performed on stamp-sized, polyamide thin-layer sheets (Chang and Creaser 1977); however, HPLC systems for additional confirmation of the identified amino acids and their quantitation have been developed (Chang 1988; Lehmann and Wittmann-Liebold 1984).

10.2
Description of the Method

Equipment   The manual procedures described here requires no expensive equipment; however, the following equipment facilitates sensitive sequencing and easy performance. The best suited set-up is a laboratory table, e.g., of 1 m length, within a hood. It should be equipped with the following items:

- nitrogen tank connected to a pipet by a plastic outlet line closeable by a stopcock

- tabletop centrifuge with holders suitable for Edman tubes (e.g. Labofuge I, Christ, Heraus,Osterode, Germany)

- vacuum device (pump, cool trap or KOH filter, manifold for 2–6 desiccators, vacuum gauge and sensor)

- one or two aluminum heaters with holes into which the Edman tubes and conversion bottles fit tightly

- Edman glass tubes, 4.5 cm long, i. d. 8 mm, with glass stoppers; alternatively: 2.5 cm, i.d. 2–4 mm

- conversion bottles, 5 cm, i.d. 4 mm

- plastic holders for 10–20 bottles that fit into the desiccators

- automatic pipettes (2, 5, 10, 20, 50, 100, and 200 µl)

- items for thin-layer chromatography: 2–4 glass beakers (25 ml, height 5 cm) with polished ends, glass covers with a layer of parafilm to seal the beakers, glass capillaries, polyamide sheets (2.5 × 2.5 cm), tweezers, holders for the ten 2.5 × 2.5 cm sheets, cold fan

- HPLC system for the identification of the DABTH derivatives by isocratic (Lehmann and Wittman-Liebold 1984) or gradient system(Chang 1988)

- DABITC (Fluka, Buchs, Switzerland, or Pierce) recrystallized **Solvents** from boiling acetone (Pro analysis grade, dried over molecu- **and reagents** lar sieve). Dissolve 1 g in 70 ml, pass through a paper filter, and allow to cool slowly: yield 0.7 g of brown needles, b.p. 169°–170 °C.

- trifluoroacetic acid (TFA), sequencing grade

- PITC, sequencing grade

- pyridine, pro analysis grade (Merck, Darmstadt, Germany), distilled successively over KOH pellets, ninhydrin and KOH, b.p. 114°–116 °C

- water : twice glass distilled and freshly prepared

The remaining reagents must be either (sequencing or pro analysis grade) and are used without further purification.

- n-heptane

- ethyl acetate

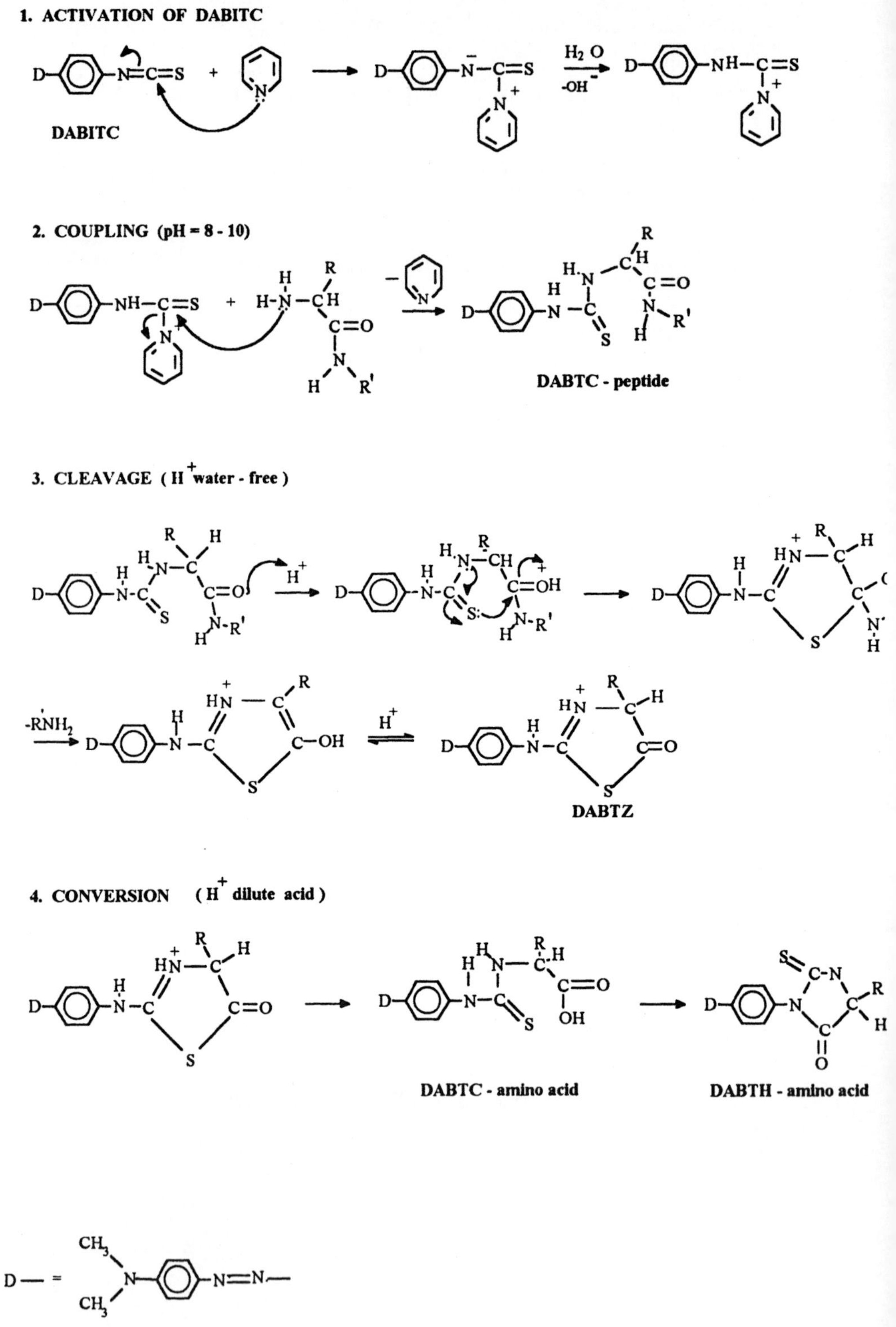
1. ACTIVATION OF DABITC
DABITC
2. COUPLING (pH = 8 - 10)
DABTC - peptide
3. CLEAVAGE (H+ water - free)
-R'NH2
DABTZ
4. CONVERSION (H+ dilute acid)
DABTC - amino acid
DABTH - amino acid
D— =

- n-butyl acetate

- acetic acid

- toluene

- n-hexane

- acetone

The reaction scheme for the degradation with DABITC/PITC is given in Fig. 1. The coupling results in the formation of a dimethylamino-azobenzene-thiocarbamyl peptide (DABITC peptide) under alkaline conditions (pH 8–10). The cleavage reaction occurs in the presence of strong anhydrous acid and releases the first amino acid as dimethylaminoazobenzene-thiazolinone derivative (DABTZ amino acid). After extraction this derivative is converted to the more stable dimethylaminoazobenzenethiohydantoin, which yields red-colored products upon treatment with HCl vapors. Reaction scheme

10.3
Experimental Procedure

10.3.1
Pretreatment of Peptide or Protein for Microsequencing

- A general cut-off of 80 mol% is recommended, but occasionally samples may require a higher purity or can be sequenced at lower purity. Purity should be verified by two separate systems (HPLC and PAGE). Polypeptide Isolation

  The sample should be salt-free to guarantee a correct pH during the coupling reaction with the DABITC/PITC reagent. Desalt the protein or peptides on C_8 or C_{18} short columns, if necessary, and dry them carefully to remove traces of acids.

- Proteins can often be precipitated from salts and undesirable buffer components (e.g., 10% glycerol) with acetone or trichloroacetic acid. This method however is not guaranteed to work.

Fig. 1. Reaction scheme for degradation with DABITC/PITC

Preparation of Sample for Degradation

1. Small peptides need no further processing as long as they are salt- and detergent-free; simply redissolve in a small amount of aqueous alcohol or dilute acid, add 5–20 µl (1000–2000 pmol) to a sequencing tube and dry the sample.

2. Large clean peptides should be redissolved in 88% formic acid, transferred to the sequencing tube and also dried.

3. Most proteins will precipitate from a concetrated aqueous solution with 1–3 volumes of acetone or from formic acid/ethylacetate, thereby freeing samples of salt or detergent.

4. Ethanol precipitation of electroeluted-electrodialyzed samples.

4a. Dissolve the lyophilized sample in 50 µl distilled water.

4b. Before procceding further, use magnification to verify that all solid material is in solution.

4c. Add 450 µl of –20 °C ethanol; mix the solution and store at –20 °C for 4–18 h.

4d. Pellet the material by centrifugation either with a swinging bucket rotor) or a fixed-angle rotor. The smaller distribution area achieved with the former improves the visibility of the precipitate.

4e. Redissolving the sample can be a problem because most of the SDS has been removed by the precipitation procedure. Recognizing that the volumes need to be kept to a minimum, add 80 µl of 50% pyridine in water and check sample solubility. If insoluble, then add 1 µl 10% SDS and recheck the solubility.

5. Transfer sample to Edman tubes shortly before use: drying is best done in the tubes used for the degradation to avoid transfer losses of sample.

6. Avoid storage of peptide under vacuum for more than several hours.

7. Remove traces of acids in the sample by repeated drying in vacuo with the addition of redistilled water or ethanol.

10.3.2
Manual Liquid Phase DABITC/PITC Double Coupling Method

The protocol is based on 2–5 nanomol peptide or protein and 500 pmol to 1 nanomol (in parentheses). The degradations are performed in small glass-stoppered tubes (Edman tubes, length 4.5 cm, i.d. 8 mm) or in shortened dansyl glass tubes (length 2.5 cm, i.d. 4 mm). For reagent and solvent purification see Sect. 10.2.2.

1. Dry peptide or protein in Edman (or dansyl) tube in desicca-   **Coupling**
   tor over KOH pellets at 100–200 millitorr); repeat drying if
   purified from acid solution.

2. Add 80 µl (20 µl) of 50% pyridine in redistilled water.

3. Flush with nitrogen or argon.

4. Add 100-fold excess of DABITC (MW 282), dissolved in pyridine, e.g., 150 µg/40 µl (30 µl/10 µl).

5. Flush with inert gas ; close tube with glass stopper (parafilm if
   using dansyl tube).

6. Incubate for 30–40 min at 52 °C in aluminum heating block.

7. Add 10 µl (2 µl) concentrated PITC.

8. Purge with inert gas, close tube.

9. Incubate for 20 min at 52 °C.

1. Add 400 µl (100 µl) n-heptane/ethyl acetate(2:1, v/v). Repeat   **Washing after**
   two to three times.   **coupling**

2. Centrifuge for 2 min in a Labofuge I (Heraeus/Christ, Oste-
   rode, Germany) tabletop centrifuge or for 2–3 s, using an Eppendorf tube as a holder, in the Microfuge B centrifuge (Beckmann, Berkeley, CA, USA). There should be clear separation
   of the phases.

3. Withdraw upper phase with a water pump and discard into a
   clean, empty waste bottle; store in case the sample is a small
   hydrophobic peptide. The withdrawal should be done with
   particular care in case of precipitation of the polypeptide at
   the interphase.

4. Dry water phase for 15–20 min in desiccator in vacuo at
   100 millitorr. Do not proceed to cleavage reaction if sample is
   not dried completely; repeat drying of the sample in vacuo
   with the addition of 40 µl (10 µl) ethanol, if necessary.

Cleavage
1. Add 50 µl (15 µl) anhydrous TFA.

2. Purge with inert gas, close with stopper.

3. Incubate for 10 min at 52 °C.

4. Dry in desiccator over KOH pellets in vacuo at 100 millitorr.

5. Repeat cleavage one to three times in case of repetitive Val-, Pro- and Ile-residues.

6. Repeat drying with the addition of 50 µl (10 µl) n-butyl acetate if sample is not dried completely.

Extraction of the thiazolinone derivative
1. Add 30 µl (30 µl) of water and 100–200 µl (30 µl) n-butyl acetate.

2. Purge with inert gas.

3. Vortex and centrifuge for 2 min in a Labofuge I tabletop centrifuge (3–5 s in a Beckman Microfuge B, Berkeley, CA, USA).

4. Withdraw upper layer containing the thiazolinone derivative into a small glass tube(length 5 cm, i.d. 4 mm) (2.5 cm, i.d. 4 mm).

5. Repeat extraction and dry the combined extracts in vacuo.

6. Dry water phase containing the residual peptide in vacuo over KOH pellets at 100 millitorr and begin next degradation cycle.

Conversion to DABTH amino acid derivative
1. Add to dried butyl acetate extract 40 µl (10 µl) 40%TFA in water.

2. Purge with inert gas ; close with parafilm.

3. Incubate for 30 min at 52 °C.

4. Dry in vacuo.

Identification
Separation of DABTH amino acid derivatives on two-dimensional polyamide thin sheets is shown in Fig. 2.

1. Dissolve DABTH amino acid in 2 µl (< 0.5 µl) of ethanol.

2. Apply approximately 1/20 (1/2 to 1/5) onto 2.5 x 2.5 cm polyamide sheets, spot size of 1 mm ; add one droplet of markers (see below).

3. Develop sheets in two dimensions: first dimension: 33% acetic acid in water ; second dimension: toluene/n-hexane/acetic acid (2:1:1, v/v).

Note: add the second dimension solvent to the beaker immediately before use.

4. Repeat sample application with appropriate amounts; comigrate the identified DABTH amino acid with the corresponding reference DABTH amino acid.

5. Inject remainder of the DABTH amino acid into a reversed phase C18 HPLC column for quantification or if any doubts exist regarding the identity of the DABTH amino acid, e.g., differentiation of DABTH-Ile/Leu (Fig. 3).

1. Pipet 500 µl of 50% pyridine into glass-stoppered test tube.   **Preparation of markers**

2. Add 30 µl each of diethylamine and ethanolamine (both redistilled).

3. Flush with nitrogen.

4. Add 250 µl DABITC in 100% pyridine (2.3 mg/ml pyridine).

5. Flush with nitrogen, close with stopper.

6. Incubate for 1 h at 52 °C.

7. Dry in vacuo.

8. Dissolve in 1 ml ethanol.

9. Apply one small droplet onto polyamide sheet together with sample. The two blue marker spots should be visible after chromatography as only very faint spots.

10.3.3
Identification of DABTH Amino Acid Derivatives

1. Cut polyamide thin-layer sheets (purchased from Schleicher and Schuell) to sizes of 2.5 x 2.5 cm using a paper cutter and keep the sheets in closed boxes. Use self-made thin glass capillaries for sample application.   **Thin-layer technique**

2. Add marker samples (see above) as smallest possible spot onto sheet (at 4 mm distance from edges).

3. Dissolve DABTH derivative in 2–5 µl of ethanol and spot portions on top of marker.

4. Develop sheet for about 4–5 min in the first dimension (33% acetic acid in water); remove the TLC sheet from the beaker when the solvent front is 2 mm from the top of the sheet. Dry with cold fan for 5 min.

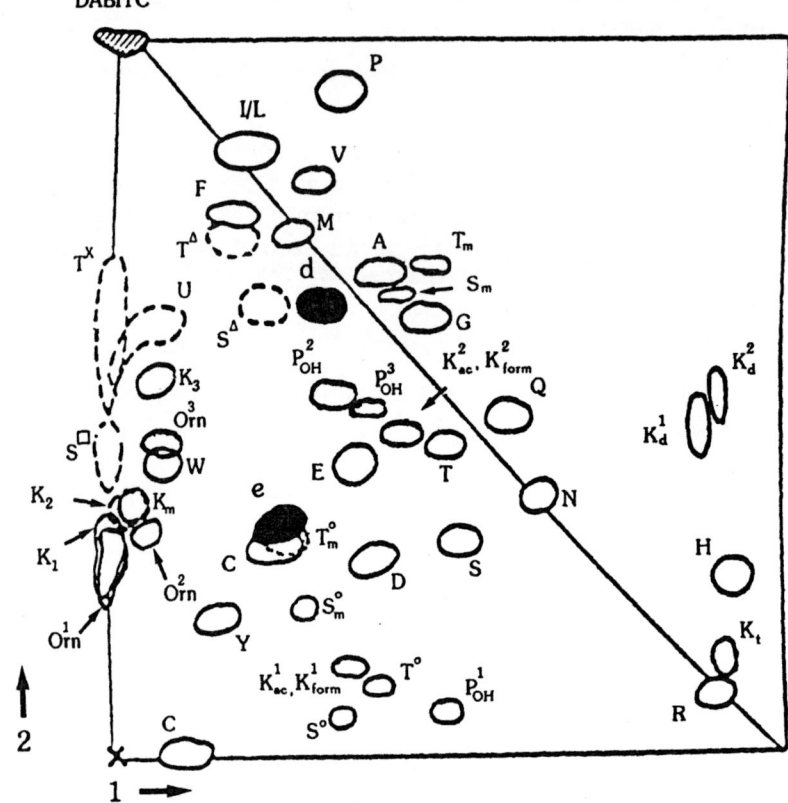

5. Develop sheet for 2 min in the second dimension (toluene:n-hexane:acetic acid, 2:1:1, v/v); remove when the front is 4–5 mm from the top. Dry by cold fan.

6. Keep polyamide sheet over HCl vapors (12 M HCl) under hood. Identify red-colored spots in comparison to reference DABTH amino acid mixture and determine migration relative to blue-colored markers (see Fig. 2).

7. Identify DABTH amino acids and additional spots (see also Sect. 10.5).

Figure 2 illustrates resolution of the DABTH amino acid derivatives by chromatography on stamp-sized polyamide sheets (for details concerning the solvent mixtures for two-dimensional chromatography, see Sect. 10.5). The location of the individual derivatives is determined relative to the migra-

Fig. 2. Separation of DABTH amino acid derivatives on two-dimensional polyamide thin-layer sheets. *Symbols denote*: DABTH amino acid derivatives of: A, alanine; C, cysteic acid; Cm, carboxymethyl cysteine; D, aspartic acid; E, glutamic acid; F, phenylalanine; G, glycine; H, histidine; I, isoleucine; K1, α-DABTH-ε-DABTC-lysine(red); K2, α-PTH-ε-DABTC-lysine (blue); K3, α-DABTH-ε-PTC-lysine(blue); Kac1, α-DABTZ-Nε-acetyl-L-lysine after hydrolysis of the acetyl-group (blue); Kac2, α-DABTH-Nε-acetyl-L-lysine (red); Kform1, α-DABTZ-Nε- formyl-L-lysine, after hydrolysis of the formyl-group; Kform2, α-DABTH-Nε-formyl-L-lysine (red); Km, derivative of Nε-methyl-L-lysine; Kd1, derivative of Nε-dimethyl-L-lysine; Kd2, derivative of Nε-dimethyl-L-lysine; Kt, derivative of Nε-trimethyl-L-lysine; L, leucine; M, methionine; N, asparagine; O1, α-DABTH-ε-DABTC-ornithine; O2, α-PTH-ε-DABTC-ornithine; O3, α-DABTH-ε-PTC-ornithine (blue); $P_{OH}1$, DABTZ-4-hydroxy-proline (blue); $P_{OH}2$, 4-hydroxy-proline (red); $P_{OH}3$, 4-hydroxy-proline after hydrolysis (red); Q, glutamine; R, arginine; S, serine; S-Δ, dehydro-serine, S·, polymerization product of serine; S°, DABTZ-serine (blue); S_m, o-methyl-L-serine; S_m°, DABTZ-o-methyl-serine (blue); T, threonine; TΔ, dehydro-threonine; T°, DABTZ-threonine (blue); Tx, blue-colored polymerization product of threonine; T_m, o-methyl-L-threonine; T_m°, DABTZ-o-methyl-L-threonine (blue); U, thiourea derivative; V, valine; W, tryptophan; Y, tyrosine; d and e, blue-colored reference markers of DABTC-reacted diethylamine and ethanolamine

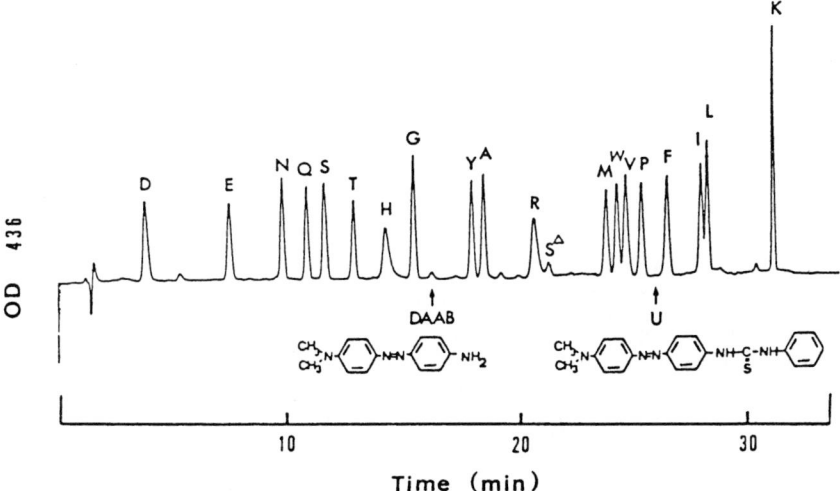

Fig. 3. Chromatogram for the complete separation of DABTH amino acids according to Chang (1988). Chromatographic conditions are as follows: column, spherisorb ODS-2.5 µm, 4.6mm×12.5 cm. Solvent A was 35 mM sodium acetate buffer (pH 5.0). Solvent B was acetonitrile . The gradient was 36% to 60% B in 25 min, 60% to 80% B in 25 to 26 min, held at 80% B from 26 to 32 min, and then back from 80% to 36% B during 32–34 min. The detector was set at 436 nm, 0.004 AUFS

tion of the two markers, DABITC-reacted diethylamine and ethanolamine (see above). Whereas the released DABTH amino acid derivatives yield red-colored spots upon exposure to acid vapors, these marker compounds result in blue spots only. The detection limit for the DABTH amino acids on micropolyamide sheets is about 20 picomol.

HPLC separation of DABTH amino acid derivatives   The released DABTH amino acids can be additionally identified by HPLC using the gradient system described by Chang (1988). The column is ODS-2, 5 μm, 4.6 mm x 12.5 cm. Solvent A is 35 mM sodium acetate buffer (pH 5.0), and solvent B acetonitrile. The applied gradient is 36 to 60% B in 25 min, 60% to 80% B from 25 to 26 min, held at 80% B from 26 to 32 min and then back from 80% B during 32 to 34 min. The detection takes place at 436 nm, 0.004 AUFS.

10.4
Problem Solving

Salts generally should be avoided because they lower the coupling yields of the polypeptides with isothiocyanates (the pH of the coupling reaction should be adjusted to pH 9.0). A lower pH hinders sufficient coupling yields mainly for histidine and lysine residues. Further salt contaminations of the peptide cause slower drying of the sample after the coupling and cleavage stages. Since salts interfere with an optimal resolution of the amino acid derivatives on thin layer-sheets, mainly in the area of the basic DABTH derivatives DABTH-Arg and -His, it is necessary to start with salt-free polypeptide preparations for the degradation and to ensure complete removal of both the base and the acid after the coupling and cleavage stages of the reaction. The derivatives of leucine and isoleucine cannot be differentiated on polyamide thin-layer chromatography; however, they separate well by the HPLC gradient system which is recommended in this chapter.

The dehydrated forms of DABTH-Ser and -Thr derivatives are formed at the cleavage step by the addition of TFA, which is a very strong acid, and at the drying stages of the Edman degradation. These additional derivatives can be diminished by choosing the appropriate conversion medium, but, no conversion medium is available which is optimal for all the different thiazolinones. Whereas 1 M HCl is quite appropriate for the conversion of serine and threonine thiazolinones, dilute TFA is more suitable for isomerization of the asparagine and glutamine derivatives. These compounds would yield considerable amounts of the free acid thiohydantoins by conversion in HCl. It may be advisable to change from one conversion medium to another for different peptides, depending on their amino acid composition, or to perform a second degradation of the same peptide employing another conversion medium.

10.5
Troubleshooting

• Reagent stability

All reactions must be maintained under nitrogen for the following reason: Once coupling has occurred, the DABTC protein is susceptible to desulfurization. Replacement of sulfur with oxygen from the atmosphere inhibits cyclization and cleavage, thus reducing yields. PITC can degrade with prolonged exposure to light and has to be stored under nitrogen or argon at –20 °C.

Owing to the limited solubility of DABITC in organic solvents, which prevents arbitrary increases of the molar ratio of DABITC peptide, the quantitative coupling of the NH_2-terminal group with DABITC can only be achieved under rather drastic conditions (75 °C, 1.5 h). This high temperature promotes rapid hydrolysis of DABITC and could cause some unexpected side reactions. Use of the DABITC-PITC coupling method enables quantitative coupling to be carried out at a moderate temperature (52°–55 °C). DABITC is not very stable in pyridine and is best prepared freshly daily or at every degradation cycle.

• Presence of salts

Samples recovered from SDS gels by electroelution-electrodialysis and other procedures usually need to be freed of contaminating salts and stain prior to sequencing. This is easily done by utilizing a precipitation method that relies on the insolubility of proteins at –20 °C in high concentrations of organic solvents (see "Preparation of Sample for Degradation," above). The presence of salts causes spot broadening in the areas of DABTH-Arg and DABTH-His: Dry the DABTH amino acid under nitrogen, add 50 µl water and reextract the derivative with butyl acetate; the salts will stay with the water phase.

• Extraction steps

In the extraction steps, unwanted materials that could hinder later reactions or form by-products which could interfere with the chromatographic analysis of DABTH amino acids are removed. The heptane:ethyl acetate mixture removes most of the unwanted materials. Nonetheless, excessive extraction can cause extraction loss of some small peptides or hydrophobic peptides. Watch the peptide layer between the organic and water phases, which is formed by precipitation of hydrophobic peptides. Remove upper layer with care; do not touch the precipitate with pipet.

Two extractions with heptane:ethyl acetate washes out 97%–98% of the excess reagents and by-products. Nevertheless, due to incomplete removal of the trace amount of the organic phase (after the second extraction), which might interfere with the later identification of DABTH amino acids, a third washing is also recommended when the yields of DABTH amino acids become very low. In this third extraction step the COOH-terminal lysine may get lost; however the amino acid can be recognized from the red color of the aqueous phase after extraction of DABTZ amino acids.

- Identification of DABTH amino acids

If "double spots" occur in the polyamide chromatography, this means that conversion of thiazolinone to thiohydantoin derivative is not complete. In this case, check the temperature of the aluminum heating block and make sure the reagent tube fits tightly into the block. If necessary, use aluminum foil for a better fit of the test tube into the block holes.

If the problem persists, prolong the conversion time, raise the temperature, or change conversion medium (20% TFA in water; 40% TFA in water; 1 M HCl; 40% TFA in ethyl acetate; methanolic HCl; acetic acid saturated with HCl gas).

The by-products of the amino acid derivatives are thought to be derived from the intermediates, possibly related to (or being) the DABTC and DABTZ derivatives. Both of these have been reported to appear blue on polyamide sheets after HCl exposure. In spite of the variations which may to some degree be dependent on chemicals and conditions, secondary spots are usually observed to some extent in the routine degradations (Bahr-Lindstrom et al. 1982).

- Difficulties with identification of the hydrophobic amino acid derivatives (DABTH-Ile/Leu; -Val/Phe)

Leucine and isoleucine can be distinguished on silica plates or by HPLC (see "HPLC Separation of DABTH Amino Acids," above). In addition, due to their different by-products, according to (Bahr-Lindstrom et al. 1982) they can also be distinguished on the polyamide sheets. Use the above described methodology for identification of the Val and Phe DABTH derivatives.

- Identification problems with DABTH-Ser and DABTH-Thr
  Apply 1 M HCl or HCl-saturated acetic acid as conversion medium.

- Double spots in case of DABTH-Asn and DABTH-Gln, caused by deamidation

Use 20%–40% TFA in water as the conversion medium to avoid formation of the acids.

- Identification of extra spots, including some posttranslationally modified amino acids (Tsiboli et al., manuscript in preparation; see Fig. 2).

10.6
Conclusions

The DABITC/PITC sequencing method is simple, sensitive and inexpensive. One degradation cycle takes ~140 min, needs no special instrumentation or radioactive materials and thus should provide a generally applicable method for micro-sequence analysis of peptides and proteins. A single polyamide TLC identification is adequate to distinguish all the common amino acids as well as some modified ones (as their DABTH derivatives), except for leucine and isoleucine (see Sect. 10.5). Identification is facilitated by the red DABTH amino acid derivatives, as well as their sometimes blue or purple by-product derivatives.

In contrast to the dansyl-Edman technique, the method is applicable to both large proteins and small peptides; thus purity checks of isolated protein and peptides, e.g., after HPLC purification, is possible.

References

Bahr-Lindstrom H, Hempel J, Jornvall H (1982) J Prot Chem 1: 257–262
Chang JY (1988) Anal Biochem 170: 542–556
Chang JY, Creaser EH, Bentley KW (1976) Biochem J 153: 607–611
Chang JY, Creaser EH (1977) J Chromatogr 132: 303–307
Chang JY, Braner D, Wittmann-Liebold B (1978) FEBS Lett 93: 205–214
Edman P, Begg G (1967) Eur J Biochem 1: 80–91
Edman P, Henschen A (1975) In: Needleman SB (ed) Protein sequence determination. Springer, Berlin Heidelberg New York, pp 232–279
Lehmann A, Wittmann-Liebold B (1984) FEBS Lett 176: 360–364
Tsiboli P, Konstantinidis G, Katsani K, Skendros Y, Choli T (1996) (manuscript in preparation)
Wittmann-Liebold B, Kimura M (1984) In: Walker JM (ed) Methods in molecular biology, vol 1. Humana, Clifton, NJ, pp 221–242

Chapter 11

Solid-Phase Sequencing of Peptides and Proteins

J. SALNIKOW

11.1
Background

The Edman degradation represents the classical reaction for chemical sequencing of peptides and proteins. In liquid-phase sequencing this reaction is homogeneous; separation of excess reagent and product is achieved by exploitation of the different partition coefficients of two immiscible phases (water/organic). This principle was perfected in the first spinning cup sequencer, designed by Pehr Edman himself. In modern gas-phase sequencers the physical conditions are similar except that the sample is fixed by noncovalent forces to a solid matrix (e.g., Immobilon, Millipore, Eschborn, Germany) and reagents are applied in liquid- or gas -phases. Miniaturization of the whole system in combination with microvolume dosage devices yielded the high sensitivity typical of these instruments.

The partition coefficients of hydrophobic peptides are not too distinct from those of the reagents. Nonetheless, a gradual loss of sample by wash-out has often been observed. This limitation prompted Laursen, in 1971, to propose covalent linkage of the protein/peptide to a solid matrix for sequencing – the principle behind solid-phase sequencing. Covalent attachment of the sample permits any excess of reagent as well as a wide variety of solvents with easy removal of all chemicals by simple washing. At the same time sample loss by wash-out is avoided; thus, solid-phase sequencing offers the highest flexibility available in Edman chemistry. Its main advantages are the following:

- There is no wash-out of sample.

- Polar solvents are tolerated, e.g., for the removal of very polar phenylthiohydantoins (PTHs) such as those derived from phosphorylated amino acids.

- New Edman reagents with easily detectable chromophores/fluorophores can be applied.

Solid-phase sequencing, in particular, has permitted evaluation of promising new Edman–type reagents in a double-coupling protocol used in conjunction with phenylisothiocyanate (Salnikow et al. 1981, 1987; Salnikow 1986, 1992).

Problems related to solid-phase sequencing are a result of the chemical attachment procedures. There are only a few chemical strategies of practical value. The most desirable attachment – an exclusive COOH-terminal linkage – is not a general method and is possible only in a few special cases. General methods lead to multi-point immobilization, which means that, in case the last amino acid is not immobilized, the polypeptide portion following the last linkage will fall off the matrix and will no longer be available for sequencing. Furthermore, the attachment yields may be variable. As a consequence of virtually quantitative attachment, the amino acids providing the linkage side chains will stay bound as PTH derivatives to the matrix and will be released only in minute quantities or not at all. However, these amino acids are more easily identified during sequencing when the immobilization yield is far from completion. Unfortunately, a useful compromise between attachment yield and identification of immobilized amino acids is experimentally rather haphazard; in most cases, therefore, quantitative attachment yields of small peptide/protein samples will be preferred, sacrificing the unequivocal identification of all amino acid residues including the matrix-linked ones.

11.2
Principles

Currently, with an amino group-containing matrix, three different attachment approaches can be used for solid-phase sequencing.

11.2.1
Immobilization with DITC (p–phenylene diisothiocyanate)

DITC is a bifunctional Edman reagent. In the first step, the amino group carrying matrix is activated forming a polymeric Edman derivative, which in the second, coupling, step leads to immobilization of the protein/peptide sample via the NH_2-terminal and additional lysyl side chains (Fig. 1). During Edman degradation of the immobilized sample the NH_2-terminal residue will be released in the cleavage step exposing the next amino acid ready for the second cycle (Fig. 2). For proteins, multi-point attachment is the rule; the polypeptide chain will stay bound to the

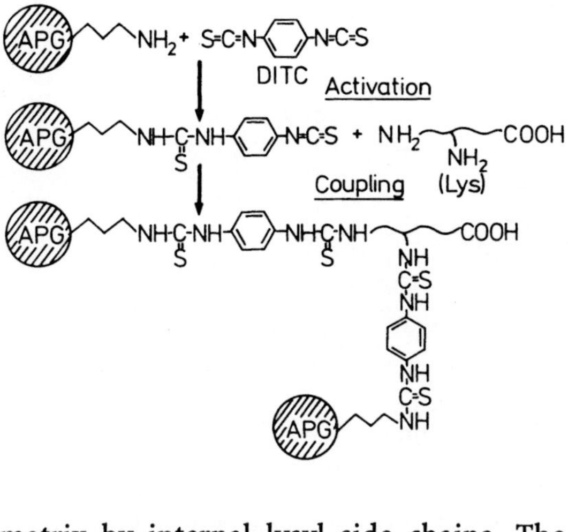

Fig. 1. DITC attachment of proteins/peptides to an amino group carrying matrix (*APG*, aminopropyl glass)

matrix by internal lysyl side chains. The NH$_2$-terminal amino acid and internal lysine residues can sometimes be identified if the attachment yield is not quantitative. Tryptic peptides with COOH terminal lysines can be sequenced up to the penultimate residue; other peptides possessing internal lysyl side chains will be released from the matrix after the last covalent linkage. Peptides which do not contain lysines are attached only via the NH$_2$-terminal and cannot be sequenced.

11.2.2
Immobilization with EDC
(N-ethyl-N'-3-dimethylaminopropyl carbodiimide)

An alternative to DITC attachment is coupling via carboxyl groups yielding again a multi-point linkage if the sample contains additional carboxyl side chains. The carboxyl attachment, however, always implies the desirable COOH-terminal fixation, permitting sequencing up to the last amino acid (Fig. 3). For efficient carboxyl attachment the nature of the applied matrix is very important: whereas aliphatic amino groups lead to poor immobilization yields, aminophenyl resins are a better choice

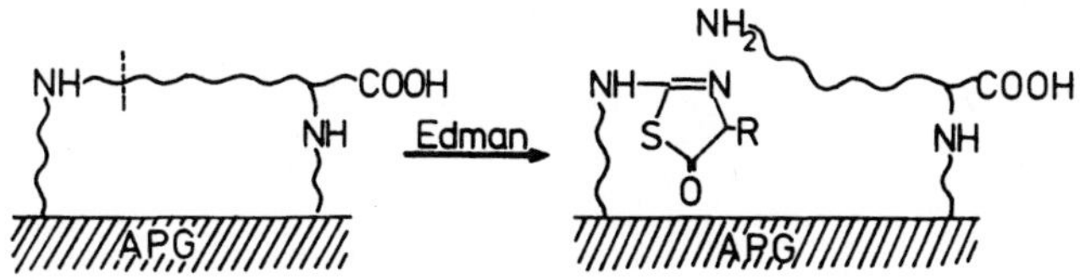

Fig. 2. Cleavage of the NH$_2$-terminal amino acid during the first Edman degradation cycle

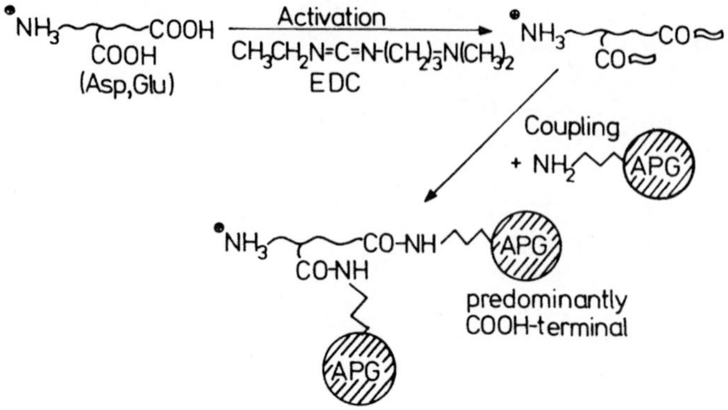

Fig. 3. Immobilization with EDC (*APG*, aminopropyl glass)

because of the higher nucleophilicity of arylamines at pH 5 (Liang and Laursen 1990).

11.2.3
Lactone Immobilization

An exclusive attachment via the COOH-terminal amino acid is possible when its carboxyl group is selectively activated. This is the case with lactone formation as observed during cyanogen bromide cleavage of polypeptides containing methionine (Fig. 4). An analogous reactive derivative is formed when a tryptophanyl bond is cleaved by treatment with dimethyl sulfoxide/halo acid or iodosobenzoic acid (Mahoney and Hermodson 1979; Mahoney et al. 1981).

These reactions are extremely useful for solid-phase sequencing, yielding the optimal single-point attachment at the COOH-terminal, at the same time sequencing up to the last amino acid is possible. This type of attachment is, however, restricted to polypeptides possessing methionine and/or tryptophan.

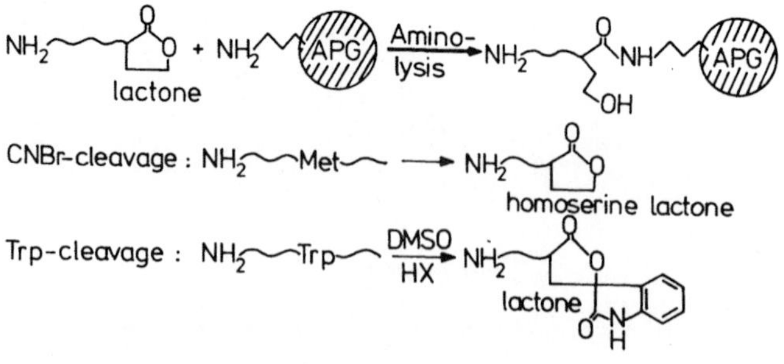

Fig. 4. Lactone immobilization (*APG*, aminopropyl glass)

The different attachment methods were developed originally for a solid-phase technology using reaction columns filled with the appropriate glass bead derivatives. Although manual degradation protocols have been described (Chang 1979; Rodrigues et al.1994), the majority of solid-phase sequencing technologies are automated processes. The availability of gas-phase sequencers with higher sensitivity has initiated a renaissance of automated solid-phase attachment chemistry using either glass filters or polyvinylidene difluoride (PVDF) membranes. This strategy has, in addition, the benefit that sequencing can directly be linked to high efficiency gel electrophoresis via modern blotting techniques, circumventing laborious and loss-bound elution procedures.

11.3
Methods

Matrices used for solid-phase sequencing consist of functionalized polystyrene, glass, or PVDF (Immobilon). Polystyrene, however, is not favored anymore because of its swelling properties in organic solvents, posing problems in automated column technologies. In addition, its high hydrophobicity limits the application of new Edman reagents possessing hydrophobic functions leading to strong interaction with the matrix. Aminopropyl and aminophenyl glass matrices can easily be prepared by reaction of glass with the respective triethoxy silane derivatives. In column solid-phase technology CPG (controlled pore glass), with its extended inner surface, provides a suitable matrix with sufficient binding capacity for polypeptides and excellent flow characteristics. The coupling capacity can be determined by Schiff base reaction with 2-hydroxynaphthaldehyde and spectrophotometric determination of the aromatic aldehyde after its release from the matrix (Schmitt and Walker 1977).

For gas-phase sequencers using membrane or glass filter disks, amino group carrying derivatives are also available. Arylamine glass filters can be prepared in a manner analogous to that used for CPG. PVDF membranes carrying arylamine side chains as well as DITC-modified derivatives are commercially available and are distributed under the trade name Sequelon (Sequelon-AA, arylamine, Sequelon-DITC, DITC-modified PVDF; MilliGen/Biosearch, Division of Millipore). Their rather high price prompted the development of simple chemical protocols starting from the less expensive PVDF membranes and resulting in good yields of the respective amino derivatives (Rodrigues et al.1994).

11.3.1
Glass and Membrane Matrices

Preparation of aminopropyl (or aminophenyl) glass

1. Place 4 g CPG beads in a round flask and dry them in a vacuum chamber at 180 °C for 2 h.

2. Add 30 ml of toluene saturated with Ca_2SO_4.

3. Add 3 ml 3-aminopropyl triethoxysilane (or aminophenyl triethoxysilane).

4. Degass shortly with a water pump.

5. Add a magnetic stirring bar and incubate the mixture for 20 h at 75 °C with occasional stirring at the lowest speed possible.

6. Remove the CPG beads by filtration through a glass frit.

7. Wash the CPG derivatives thoroughly with toluene and acetone.

8. Dry the CPG derivatives in a desiccator in vacuo.

Note: CPG with 75 Å pores is, in general, sufficient for most peptides and proteins. For very large polypeptides wider pore sizes are recommended; however, these have a smaller inner surface and, thus, a reduced coupling capacity.

Determination of amino groups

This protocol is based on that of Schmitt and Walker (1977).

1. Mix 10 mg aminopropyl glass in a small test tube with 2 ml 0.2 M 2-hydroxy-1-naphthaldehyde (dissolved in freshly distilled dimethyl formamide) and incubate at room temperature for 15 h with slow stirring.

2. Wash with 4 x 3 ml dimethyl formamide followed by several 3 ml portions of ethanol until the absorption at 270 nm reaches zero (measured against pure ethanol).

3. Add 2 ml 0.4 M benzylamine and incubate 45 min at room temperature with slow stirring.

4. Centrifuge and measure the absorption of the supernatant at 420 nm.

Note: The A_{420} of the benzylamine Schiff base measured in the supernatant is 10900. Good preparations of aminopropyl glass have a coupling capacity of 0.20–0.22 mmol amino groups/g glass.

This protocol is based on that of Aebersold et al. (1990).

Preparation of arylamine glass filters

1. Circular Whatman GF/F filters are kept in a glass beaker in anhydrous trifluoroacetic acid for 1 h at room temperature with occasional shaking (use an efficient hood!).

   Note: Anhydrous trifluoroacetic acid is a very corrosive and toxic chemical, use gloves and eye protection!

2. The glass filters are removed and dried on filter paper.

3. Place the preactivated glass filters in a plastic bag containing 2% aminophenyl triethoxysilane in acetone/water (1:1, v/v).

4. Incubate overnight at 37 °C by gently rotating on a shaker.

5. Wash the derivatized glass filters 10 times with acetone (5 min/wash).

6. Dry the washed glass filters for 45 min at 110 °C in an oven and store them at room temperature in a covered petri dish.

   Note: Etching of glass filters with trifluoroacetic acid is recommended to increase the reactivity towards the silane derivatives. An approximate estimation of arylamine content is possible by preparing the corresponding aminopropyl derivatives using 3-aminopropyl triethoxysilane under identical conditions and amino group determination with the 2-hydroxy-1-naphthaldehyde test (see above). Alternatively, an excess of peptide can be attached to the membrane and the coupled amount determined by hydrolysis and amino acid analysis or – in case of a radioactive sample – by scintillation counting.

This protocol is based on that of Rodrigues et al. (1994).

Preparation of amino group functionalized PVDF membranes

1. Incubate PVDF membrane disks (Immobilon-P or Immobilon-P^{SQ}) with 0.5 M KOH for 1.5 min with gentle stirring.

2. Transfer the membrane disks with tweezers to 2 ml 0.2 M Na_2HCO_3 (pH 9.0) for 1 min.

3. Transfer the membrane disks to 2 ml 50% methanol containing 2% triethylamine (v/v) and 0.1% poly(allylamine) hydrochloride (w/v) for 2 min.

4. Remove the solvent by placing the membrane disks in a desiccator under vacuum using a water jet pump.

5. Wet the dry disks again with methanol and to each disk apply a second sample (15 µl) of triethylamine/poly(allylamine) reagent (see above) and dry again.

6. Place the semi-dry disks in a vial containing 1 ml 0.2 M Na_2HCO_3 (pH 10.5) and incubate overnight.

7. Wash the disks with deionized water, followed by methanol, and then water again (3 ml, 1–2 min each time). Store the disks in methanol at –20 °C.

Note: The amino group functionalized PVDF membranes have a lower coupling capacity than commercial ones; nevertheless, they can be used for DITC coupling and automated and manual sequencing. The coupling capacity can be determined by attachment of peptide samples and amino acid analysis after acid hydrolysis (Rodrigues et al.).

Preparation of DITC glass beads (DITC-modified aminopropyl CPG)

1. Dissolve 100 mg DITC in 2 ml freshly distilled dimethyl formamide and add 160 mg aminopropyl glass (CPG 75 Å).

Note: Dimethyl formamide can be substituted by benzene. Although it is claimed that DITC glass beads can be stored at –20 °C, freshly prepared derivatives show the best coupling results.

2. Add 20 µl triethyl amine and incubate at 25 °C for 1 h under gentle stirring. The supernatant will become slightly yellowish.

3. Wash the DITC glass beads with 2 x 2 ml dimethyl formamide followed by 4 x 5 ml methanol and acetone.

4. Dry in a desiccator in vacuo.

Note: Sudden evacuating and aerating of CPG beads placed in a desiccator might lead to matrix losses by jumping of the beads. The problem can be alleviated by placing the glass beads in a vial sealed with parafilm and piercing the parafilm seal several times with a fine sewing needle. Alternatively, a vial with a fritted stopper can be used.

11.3.2
DITC Coupling Assay

Using DITC glass beads

This assay is designed for automated solid-phase sequencing using microcapillary columns filled with DITC glass (Liang and

Laursen 1990), but it can also be used also for other solid-phase technologies.

1. Dissolve 20–1000 pmoles of polypeptide in 30 μl coupling buffer: 0.2 M Na$_2$HPO$_4$ (pH 9.0), 1% SDS (sodium dodecyl sulfate).

2. Add approximately 30 μl DITC glass beads.

3. Incubate for 45 min at 55 °C.

4. Wash 2–3 times with distilled water and then methanol; dry with nitrogen.

This protocol is based on that of Herfurth et al. (1991).

Using DITC-
modified PVDF

Note: DITC coupling results usually in high to quantitative attachment yields. Whereas with DITC-modified aminopropyl glass the NH$_2$-terminal amino acid including internal lysines can sometimes be identified in low yield, with DITC-PVDF membranes the immobilization is in general quantitative. Moderate SDS concentrations are tolerated making this attachment technique the method of choice for insoluble polypeptides such as membrane proteins.

1. Dissolve 100–500 pmoles of lysine containing polypeptides in 20–40 μl coupling buffer: 0.2 M N-ethyl morpholine(pH 8.0), containing 50% ethanol, or 0.1 M NaHCO$_3$ (pH 11.4), containing 50% ethanol and 0.25% SDS

2. Apply a 5 μl portion of the polypeptide solution to the activated membrane placed in the reaction cartridge of the sequencer.

3. Dry under a stream of nitrogen.

4. Repeat the procedure with additional 5 μl portions until the whole sample is applied (total time approximately 45 min).

Note: Buffers or reagents containing amines or ammonium ions have to be avoided, since they will reduce the binding capacity of DITC matrices by competitive reaction. Attachment buffers for DITC-PVDF membranes must contain at least 20% of a water-miscible organic solvent since DITC-PVDF membranes do not wet with water (Herfurth et al. 1991). The solubility of polypeptides to be sequenced in these buffers has to be determined. Washing solvents such as methanol should be free of aldehydes to avoid Schiff base blocking of the NH$_2$-

terminal. Peptides containing no internal or COOH-terminal lysines do not sequence!

11.3.3
EDC Coupling Assay

Using amino-
phenyl glass

This protocol is based on those of Liang and Laursen (1990) and Aebersold et al. (1990).

1. Dissolve 20–1000 pmoles of polypeptide in 30 µl coupling buffer: 0.2–1 M pyridine- HCl (pH 4.0–5.0), or: 0.2–1 M Mes (2–(4–morpholino) ethanesulfonic acid; pH 3.5–5.5).

   Note: SDS is tolerated up to 1%.

2. Add EDC (0.2–2 mg dissolved in a minimum amount of water).

3. **Immediately** add 30 µl aminophenyl glass.

4. Incubate for 15–60 min (25° or 37 °C).

5. Wash with water, methanol and dry with nitrogen.

Using arylamine
glass filters

This protocol is based on that of Aebersold et al. (1990).

1. Apply peptide (100–500 pmoles in aequeous solution or in water/acetonitrile/0.1% trifluoroacetic acid from HPLC) in small portions to the arylamine glass filter disk.

2. Dry under a stream of nitrogen.

3. Apply 30 µl of coupling buffer: 0.2–1 M Mes (2–(4–morpholino) ethanesulfonic acid; pH 3.5–5.5).

4. Apply 10 µl of freshly prepared EDC (20 mg/ml water, w/v).

5. Incubate for 30–60 min at 37 °C.

6. Wash the glass filter disk with distilled water.

7. Insert the disk into the sequencer.

Using arylamine
PVDF
membranes

This protocol is based on that of Pappin et al. (1989).

1. Apply 100–300 pmoles peptide in 30%–50% acetone to the membrane disk.

2. Heat the membrane on a heating block at 55°C until dry.

3. Dissolve 1.0 mg EDC in 100 µl coupling buffer: 0.1 M Mes (=2–(4–morpholino) ethanesulfonic acid; pH 5.0), containing 15% acetonitrile.

Note: Buffers containing acetate or other components with carboxyl group functions must be avoided because they will quench the coupling reaction competitively.

4. Remove the membrane disk from the heating block and add 5 µl EDC containing coupling buffer to the surface.

5. Allow the reaction to proceed for 20 min at room temperature and accommodate the membrane disk in the sequencer cartridge.

Note: Optimal carboxyl attachment occurs at pH 3–5. Activation of the polypeptide carboxyl groups must be followed by immediate exposure to the arylamine matrix in order to warrant high attachment yields. Although the COOH-terminal of the polypeptide with its lower pK value is attached preferentially, internal Glu and Asp are also anchored, resulting in less release of the corresponding PTHs during sequencing. SDS is tolerated, making this attachment method suitable for difficult to solubilize polypeptides such as membrane proteins. For peptides lacking lysines, EDC coupling is a true alternative to DITC coupling, with the additional advantage that the COOH-terminal amino acid can be identified. EDC coupling is, however, in general less quantitative than DITC attachment.

11.3.4
Coupling of Peptides via COOH-Terminal Homoserine Lactone

1. Dissolve the dried peptide in 50 µl anhydrous trifluoroacetic acid and incubate at room temperature for 30 min.

2. Remove the acid in a stream of nitrogen (use a fume hood!) and dissolve the residue in 50 µl freshly distilled dimethyl formamide.

3. Add 20 mg of aminopropyl glass.

4. Add 10 µl triethylamine, sonicate gently and incubate for 90 min at 55 °C.

5. Wash the glass with 3 x 2 ml methanol, acetone and dry in a vacuum desiccator.

Note: Treatment with anhydrous trifluoroacetic acid prior to coupling is important to achieve quantitative lactonization and solubilization of the peptide. Consequently, it is not recommended to dry the peptide thereafter in a vacuum desiccator over NaOH which might render it insoluble (Salnikow et al. 1981).

Note: Only peptides originating from cyanogen bromide cleavage of methionyl bonds or iodosobenzoic acid cleavage of tryptophanyl bonds will yield COOH-terminal γ-hydroxy acid derivatives which will form γ-lactones. All other peptides will not be coupled with this procedure and will not sequence!

References

Aebersold R, Pipes G D, Wettenhall R E H, Nika H, Hood L E (1990) Covalent attachment of peptides for high sensitivity solid-phase sequence analysis. Anal Biochem 187:56–65.

Chang J Y (1979) Manual solid phase sequence analysis of polypeptides using 4-N,N-dimethylaminoazobenzene 4-isothiocyanate. Biochim Biophys Acta 578:188–195

Herfurth E, Pilling U, Wittmann-Liebold B (1991) Microsequencing of proteins and peptides in the Knauer Sequencer with and without covalent attachment to polyvinylidene difluoride membranes by the wet-phase degradation technique. Biol Chem Hoppe-Seyler 372:351–361

Laursen R (1971) Solid phase Edman degradation. An automatic peptide sequencer. Eur J Biochem 20:89–102

Liang S-P, Laursen R (1990) Covalent Immobilization of proteins and peptides for solid-phase sequencing using prepacked capillary columns. Anal Biochem 188:366–373

Mahoney W C, Hermodson M A (1979) High–yield cleavage of tryptophanyl bonds by o-iodosobenzoic acid. Biochemistry 18:3810–3814

Mahoney W C, Smith P K, Hermodson M A (1981) Fragmentation of proteins with o-iodosobenzoic acid: chemical mechanism and identification of o-iodosobenzoic acid as a reactive contaminant that modifies tyrosyl residues. Biochemistry 20:443–448

Pappin D J C, Coull, J M, Köster, H (1989) Covalent sequence analysis of proteins electroblotted or spotted onto PVDF membranes. The Third Symposium of the Protein Society, July 29–August 2, Univ. of Washington, Seattle, WA., MilliGen/Biosearch, Covalent Protein Sequencing Forum

Rodrigues J A, Combrink J, Brandt W F (1994) Derivatization of polyvinylidene difluoride membranes for solid–phase sequence analysis of a phosphorylated sea urchin embryo histone H1 peptide. Anal Biochem 216:365–372

Salnikow J, Lehmann A, Wittmann-Liebold B (1981) Improved automated solid-phase microsequencing of peptides using DABITC. Anal Biochem 117:433–442

Salnikow J (1986) Automated solid-phase microsequencing using DABITC, on column immobilization of proteins. In: Wittmann-Liebold B, Salnikow J, Erdmann V A (eds) Advanced methods in protein microsequence analysis. Springer, Berlin, Heidelberg, New York, pp 108–116

Salnikow J, Palacz Z, Wittmann-Liebold B (1987) Automated solid-phase sequencing using fluorescent Edman reagents. In: Walsh K A,(ed) Methods in protein sequence analysis. Humana,Clifton, NJ, pp 247–260

Salnikow J (1992) Proteinchemical methods as tools in modern biotechnology. In: Vardar-Sukan F and Sukan S S (eds) Recent advances in biotechnology. Kluwer Academic, Netherlands, pp 389–395

Schmitt H W, Walker J E (1977) Coupling capacity of solid-phase supports. FEBS Lett 81:403–404

Sequence Analysis of the NH₂-Terminally Blocked Proteins Immobilized on PVDF Membranes from Polyacrylamide Gels

H. Hirano

12.1
Background

In 1985, a new protein preparation method for microsequencing was established (Vandekerckhove et al. 1985), in which subnanomole amounts of proteins were separated by one-dimensional or two-dimensional (2D) polyacrylamide gel electrophoresis (PAGE) and then transferred from the gel onto a polybrene-treated glass fiber filter by electroblotting and sequenced directly by a gas-phase sequencer. Two years later, it was found that a polyvinylidene difluoride (PVDF) membrane could be used in place of the glass fiber filter for efficient electroblotting and gas-phase sequencing (Matsudaira 1987). The PVDF membrane filter has several advantages over the glass fiber filter in, for example, protein binding capacity, protein detection on the filter, and handling. Since 1987, electroblotting/sequencing using the PVDF membrane has come to be widely used in protein sequence analysis.

However, even if proteins are successfully separated by PAGE and transferred onto a PVDF membrane by electroblotting, NH₂-terminal amino acid sequences of the proteins with a blocking group at the NH₂-terminal cannot be determined by Edman degradation in the sequencer. Many proteins are NH₂-terminally blocked (Aitken 1990; Tsunasawa and Sakiyama 1992). Brown and Roberts (1976) estimated that over 50% of soluble mammalian proteins are NH₂-terminally blocked. Thus, there is a need for a simple and rapid technique for obtaining sequence information on blocked proteins.

Most techniques proposed so far for this purpose are applicable only to obtaining the internal sequence of proteins. New techniques for chemical and enzymatic deblockings have been developed and the NH₂-terminal sequencing of proteins with an acetyl, formyl or pyroglutamyl group at the NH₂-terminal, elec-

Table 1. NH$_2$-Terminal blocking group found in proteins (from Tsunasawa and Hirano 1993)

Blocking group	Commonly affected residue
Acyl-	
Formyl	Gly, Met
Acetyl	Ser, Ala, Met, Gle, Asp, Glu, Thr, Val, Pro
Myristoyl	Gly
α-Ketoacyl	Pyruvoyl[Ser], α-Ketobutyl[Thr]
Glucronyl	Gly
Pyroglutamyl	Glu, Gln
Murein	Lys
Alkyl-	
Methyl(mono)	Met, Ala, Phe
Methyl(di)	Pro
Methyl(tri)	Ala
Glucosyl	Val:[1-deoxy, 1-(N$^{\alpha}$-Val)-fructose]

troblotted from a polyacrylamide gel onto a PVDF membrane (Tsunasawa and Hirano 1993), can be carried out.

Acetyl modification is the most prevalent of the blocking groups identified so far (Tsunasawa and Hirano 1993); formyl and pyroglutamyl groups are also frequently detected (Table 1). In this chapter, techniques for the on-membrane deblocking of proteins with these blocking groups and subsequent gas-phase sequencing are described. Techniques for PAGE and electroblotting of proteins were presented in the previous chapters.

It should be noted that proteins are NH$_2$-terminally blocked not only in vivo but also in vitro. It is possible to prevent artificial in vitro blocking generated during protein extraction, PAGE and electroblotting. The use of highly pure reagents, addition of 100 pmol thioglycolic acid as a free-radical scavenger during the extraction, electrophoresis and electroblotting buffers, or pre-electrophoresis for removing the free-radicals from the gel may help to prevent in vitro blocking (Moos et al. 1988; Ploug et al. 1992; Walsh et al. 1988). However, if proteins are blocked in vivo, a chemical or enzymatic deblocking procedure such as that described here is required to determine the NH$_2$-terminal sequence.

12.2
Methods

12.2.1
Deblocking of Proteins with NH$_2$-Terminal Acetylserine and Acetylthreonine

Proteins with either acetylserine or acetylthreonine can be deblocked by trifluoroacetic acid (TFA) treatment (Wellner et al. 1990). About 36% and 4% of acetylated proteins carry acetylserine and acetylthreonine, respectively, as NH$_2$-terminal amino acids. Treatment with TFA vapor should thus deblock about 40% of acetylated proteins. However, proteins with other NH$_2$-acetylamino acids cannot be deblocked by this treatment. These proteins should be deblocked with acylamino acid releasing enzyme (AARE), as described below.

1. Separate the NH$_2$-terminally blocked protein (< 200 pmol) by   **Procedure**
   SDS-PAGE or 2D-PAGE and electoroblot it on a PVDF membrane.

2. Cut out the portion of the PVDF membrane carrying the protein band or spot and place the membrane in an Eppendorf tube.

3. Place the Eppendorf tube in a vial (2.6 x 6 cm) and add 100–300 µl of TFA.

4. Purge the vial with nitrogen gas for 30 s, seal with a stopper and incubate at 60 °C for 30 min.

5. Dry the membrane in vacuo and apply to a gas-phase sequencer.

Wellner et al. (1990) indicate that an N → O acyl shift may be   **Comments**
involved in the removal of the acetyl group of NH$_2$-terminal acetylserine and acetylthreonine (Fig. 1). The advantage of this method is that deblocking is easy and rapid, although overall sequencing yields obtained by this procedure are usually low (~ 5%). When the membrane is exposed to TFA vapor for longer than 2 h, yields of the PTH amino acids increase, but at the same time so do those reaction by-products which prevent identification of PTH amino acids. Reaction by-products may be generated primarily through cleavage of polypeptide with TFA (Hulmes and Pan 1991).

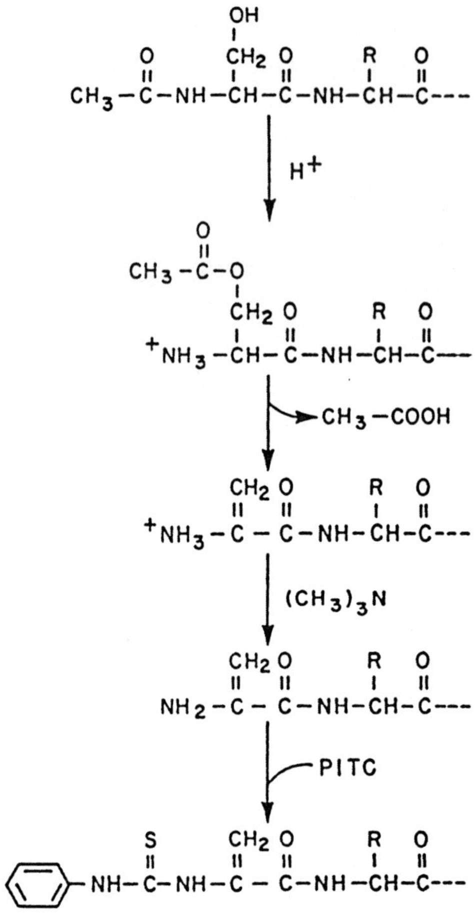

Fig. 1. Possible mechanism of deblocking reaction in N-acetylated proteins (Wellner et al. 1990)

12.2.2
Deblocking of N-formylated Proteins

The N-formyl group of proteins can be removed when the protein-blotted PVDF membrane is incubated in a dilute HCl solution (Ikeuchi and Inoue 1988).

1. Separate the N-formylated protein (< 100 pmol) by SDS-PAGE or 2D-PAGE and electroblot it on a PVDF membrane.

2. Cut out the portion of the PVDF membrane carrying the protein band or spot.

3. Wet the membrane with a small amount of acetonitrile and soak it in 200–500 µl 0.6 M HCl at 25 °C for 24 h.

4. Wash the PVDF membrane adequately with deionized water.

5. Dry the membrane in vacuo and apply it to a gas-phase sequencer.

12.2.3
Deblocking of Proteins with NH$_2$-Terminal Pyroglutamic Acid

The NH$_2$-terminal pyroglutamic acid of proteins can be removed by in situ pyroglutamyl peptidase digestion; subsequently the proteins can be sequenced from the second residue (Hirano et al. 1991; Moyer et al. 1990).

1. Separate the NH$_2$-terminally blocked protein (< 100 pmol) by SDS-PAGE or 2D-PAGE and electroblot it on a PVDF membrane.

2. Cut out the portion of the PVDF membrane carrying the protein band or spot.

3. Wet the membrane with a small amount of acetonitrile and soak it in 200 µl 0.5% (w/v) polyvinylpyrrolidone (PVP)-40 (Sigma) in 100 mM acetic acid at 37 °C for 30 min to block the unbound protein-binding site on the membrane.

4. Wash the membrane with 5 ml of deionized water at least ten times.

5. Soak the membrane in 100 µl 0.1 M phosphate buffer (pH 8.0) containing 5 mM dithiothreitol (DTT), 10 mM EDTA, and 10%(v/v) acetonitrile.

6. Add pyroglutamyl peptidase (Boehringer Mannheim) (5 µg, enzyme/substrate 1:1–1:10).

7. Incubate the reaction solution at 30 °C for 24 h.

8. Wash the membrane with deionized water.

9. Dry the membrane in vacuo and apply it to a gas-phase sequencer.

Comments

Miyatake et al. (1992) treated the protein-blotted PVDF membrane with anhydrous hydrazine vapor at 20 °C for 8 h to deblock proteins with a pyroglutamic acid at the NH$_2$ terminal. The NH$_2$-terminal pyroglutamic acid of the protein was converted to Glu-γ-hydrazide which then could undergo Edman degradation. However, this often causes partial modification of asparagine and glutamine to their hydrazides and the conversion of arginine to ornithine. Milder hydrazinolysis at –5 °C for 8 h may be useful to deblock the N-formylated residue.

12.2.4
Deblocking of N-Acetylated Proteins

The NH_2-terminal acetylamino acid of proteins dissolved in a buffer solution can be removed with AARE (Tsunasawa et al. 1990). Figure 2 diagrams the experimental procedure for deblocking of N-acetylated proteins with AARE. This technique is applicable to the deblocking of N-acetylated proteins electroblotted onto a PVDF membrane (Hirano et al. 1991).

In the case of pyroglutamyl peptidase, proteins with relatively high molecular mass such as hen egg riboflavin-binding protein (34 kDa) and *Geotrichum candidum* lipase (59 kDa) can be deblocked on a PVDF membrane by direct treatment with the en-

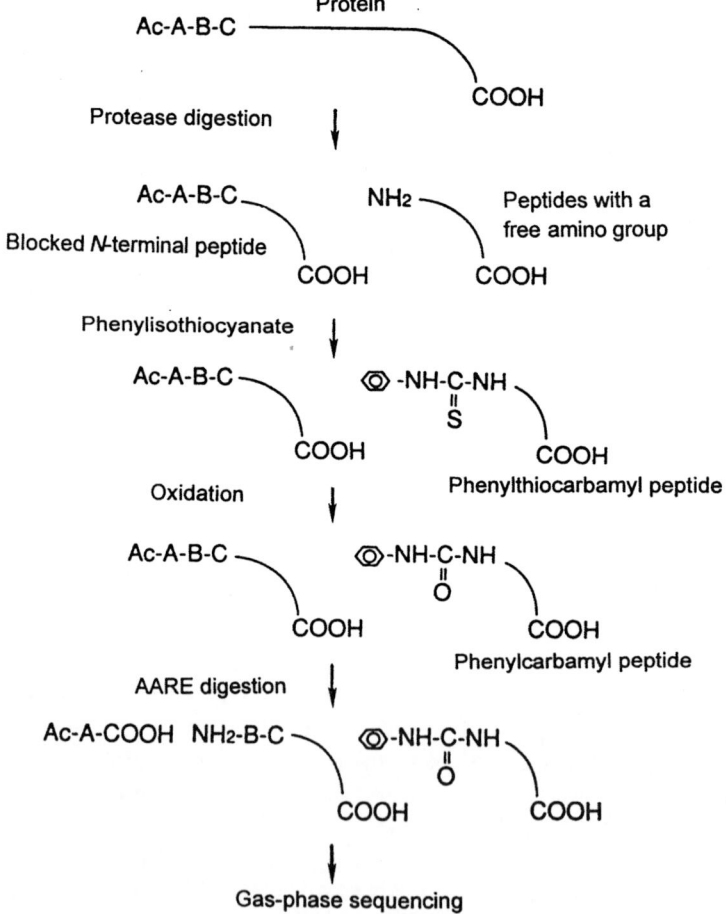

Fig. 2. Deblocking of N-acetylated proteins with acylamino acid releasing enzyme (AARE). (From Tsunasawa et al. 1990)

zyme. AARE, in contrast, does not directly digest proteins with high molecular mass. As described by Tsunasawa et al. (1990), AARE specifically cleaves the X-Y bond of RCO-X-Y type peptides shorter than 40 residues (R: alkyl group, X/Y: L-amino acid). Thus, prior to AARE digestion, proteins on the membrane should be digested with a protease such as trypsin. Short peptides produced are subsequently released from the membrane. In this protease digestion, 10% acetonitrile is included in the digestion buffer to reduce hydrophobic interaction between peptide and PVDF membrane (Aebersold et al. 1987). By addition of this organic solvent, peptides are more efficiently eluted from the membrane. The acetylated amino acid of the NH$_2$-terminal peptide extracted is selectively removed by digestion with AARE.

1. Separate the NH$_2$-terminally blocked protein (< 100 pmol) by **Procedure** SDS-PAGE or 2D-PAGE and electroblot it onto a PVDF membrane.

2. Cut out the portion of the PVDF membrane carrying the protein band or spot.

3. Wet the membrane with a small amount of acetonitrile and pretreat the membrane with 200 µl 0.5% (w/v) PVP-40 in 100 mM acetic acid at 37 °C for 30 min.

4. After thorough washing with deionized water, digest the protein on the PVDF membrane with 5–10 µg trypsin (methylated trypsin; Promega) (enzyme/substrate: 1:1–1:10) in 100 µl 0.1 M ammonium bicarbonate buffer (pH 8.0) containing 10% (v/v) acetonitrile at 37 °C for 24 h with shaking.

5. Pool the digestion buffer containing tryptic peptide in an Eppendorf tube.

6. Wash the membrane by vortexing with 100 µl deionized water and the washing solution with the digestion buffer.

7. Evaporate the digestion mixture to dryness in vacuo and add 100 µl 50% (v/v) pyridine and 10 µl PITC (sequencing grade) to bind the free amino groups of the peptides obtained.

8. Flush with nitrogen gas, vortex the reaction solution and centrifuge at 3000 g for 1 min.

9. Discard the resultant supernatant containing reaction byproducts and excess reagent. Repeat this washing procedure three times and evaporate the sample to dryness in vacuo .

10. Prepare performic acid solution by mixing 9 ml formic acid and 1 ml 30% hydrogen peroxide and keep the mixture at

room temperature for 1 h. Add 100 μl performic acid to the dried sample and, after mixing, place the tube on ice (1 h) to convert the NH$_2$-terminal phenylthiocarbamyl groups of the peptide to phenylcarbamyl groups by oxidation.

11. Evaporate the sample solution to dryness in vacuo, resuspend in deionized water and again dry in vacuo .

12. Resuspend the sample in 100 μl 0.2 M phosphate buffer (pH 7.2) containing 1 mM DTT.

13. Dissolve 50 mU AARE (Takara Shuzo) in 50 μl of the same buffer and add to the mixture. Incubate at 37°C for 12 h to remove the N-acetylated amino acid.

14. Apply the sample solution to a polybrene-coated glass fiber filter mounted into the upper glass block of the reaction chamber of a gas-phase sequencer.

Comments   • If the NH$_2$-terminal peptide obtained by protease digestion has more than 10 residues, its extraction from the PVDF membrane is difficult. In this case, a second digestion with another protease should be performed on the same membrane.

• Krishna et al. (1991) reported an alternative method in which, after fragmentation, peptides with a free amino group at the NH$_2$-terminal are succinylated and the NH$_2$-terminal blocked peptide is then deblocked with AARE.

• If butyl chloride (S3) is delivered for 30 s prior to Edman degradation in the gas-phase sequencer to recover the NH$_2$-terminal blocked amino acid released by AARE, the blocking group and NH$_2$-terminal amino acid can be identified by mass spectrometry.

• In the sequencing of cytochrome c, an N-acetylated protein, overall initial yields from successive steps including electrophoresis, electroblotting, deblocking and sequencing ranged from 23%–25%. The efficiency of deblocking and sequencing described here depends primarily on tryptic digestion and subsequent peptide elution from the PVDF membrane. Shaking the digestion solution containing the PVDF membrane during tryptic digestion should facilitate elution of the tryptic peptides.

• The NH$_2$-terminal myristoyl group of the blocked proteins can be removed when, instead of AARE, peptide N-fatty acylase (Wako Pure Chemicals) is used as described above.

12.3
Discussion

Deblocking of proteins electroblotted on a PVDF membrane has several advantages.

- Proteins purified by PAGE can be efficiently deblocked and sequenced.

- Chemical and enzymatic deblocking can be easily performed without desalting, since after electroblotting on a PVDF membrane, the proteins become almost completely salt-free.

- Proteins can be deblocked at picomole levels.

- Sequential deblocking, as described below, is possible.

The above deblocking techniques may be used in combination to allow sequential deblocking for sequencing of unknown proteins immobilized on PVDF membranes (Fig. 3).

Sequential deblocking

1. A protein sample is transferred from the PAGE gel onto a PVDF membrane by electroblotting.

2. The membrane carrying the protein is directly subjected to gas-phase sequencing.

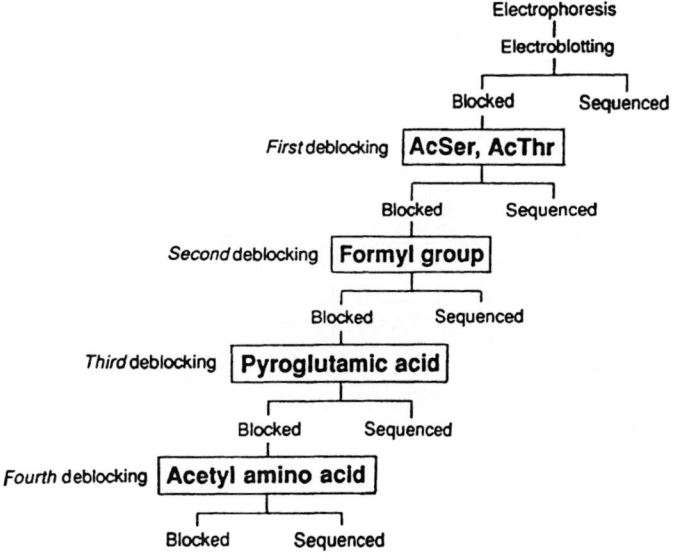

Fig. 3. Strategy of sequential deblocking for NH$_2$-terminally blocked proteins electroblotted onto a PVDF membrane

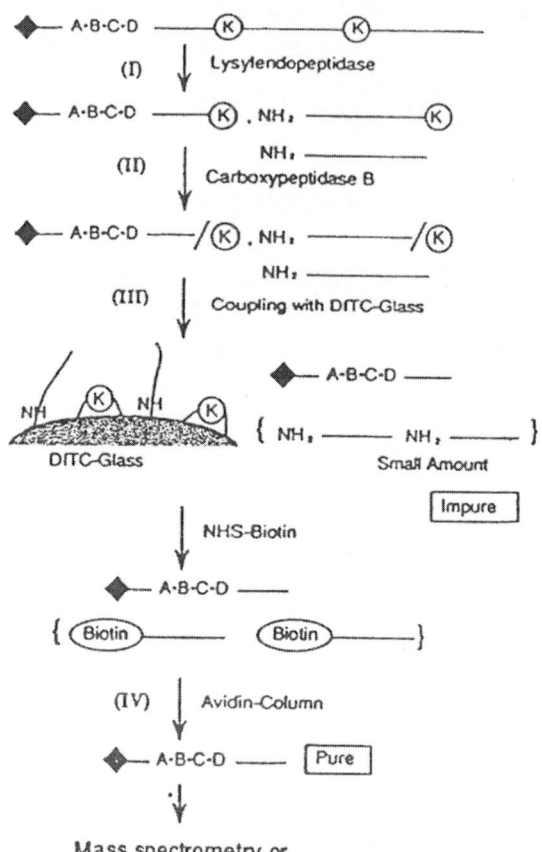

Fig. 4. Isolation of the NH$_2$-terminal peptide from NH$_2$-terminally blocked proteins (Mitsunaga et al. 1993). (1) NH$_2$-terminally blocked proteins are digested with lysylendopeptidase. (2) The resultant peptides are digested with carboxypeptidase B to remove the COOH-terminal lysine from each peptide. (3) The peptides is bound to phenylenediisothiocyanate (DITC) glass to eliminate the peptides and lysine residues except for the NH$_2$-terminal blocked peptide. (4) The unadsorbed fraction by DITC glass treatment is reacted with the N-hydroxysuccinimide-biotin to biotinylate the α and ε amino groups of the peptides and lysine residues remained. The biotinylated peptides and lysine residues are immobilized on an avidin-agarose column. The NH$_2$-terminally blocked peptide is obtained as an unadsorbed fraction

3. If sequencing (2–3 cycles) fails at step 2, the membrane is removed from the sequencer, and treated with TFA vapor at 60 °C for 30 min to remove the acetyl groups of acetylserine and acetylthreonine. If the protein is NH$_2$-terminally blocked by acetylserine or acetylthreonine, the acetyl group is removed by this procedure and sequencing from the NH$_2$-terminal becomes possible.

4. If sequencing fails at step 3, the membrane is incubated in 0.6 M HCl at 25 °C for 24 h to remove formyl groups and then resubjected to sequence analysis.

5. If sequencing fails at step4, the sample is subjected to on-membrane pyroglutamyl peptidase digestion to remove NH$_2$-terminal pyroglutamic acid and resequenced.

6. Finally, if sequencing fails at step 5, deblocking with AARE is performed to removed acetylamino acids that were not removed in step 3 and sequencing is then attempted again. If sequencing still fails, different methods must be used.

Note: When no NH$_2$-terminal sequence is obtained in Edman degradation of a certain protein, it is possible that the NH$_2$-terminal of the protein is NH$_2$-terminally blocked. However, it is impossible to know which blocking group is involved at this experimental stage. In this case, the sequential deblocking technique using proteins transferred onto PVDF membranes may be useful to identify the blocking group and actually deblock the proteins.

Otherwise, the blocked NH$_2$-terminal peptide is selectively removed (Mitsunaga et al. 1993; Akiyama et al. 1994), and the precise mass of the peptide can be determined by mass spectrometry methods, such as electrospray ionization or matrix assisted laser desorption time-of-flight, to identify the blocking group.

If the NH$_2$-terminally blocked protein is available only in picomole amounts, the biotin-avidin method recently developed by Mitsunaga et al. (1993) may be useful to remove selectively the blocked NH$_2$-terminal peptide (Fig. 4).

References

Aebersold RH, Leavitt J, Saavedra RA, Hood LE, Kent SBH (1987) Internal amino acid sequence analysis of proteins separated by one- or two-dimensional gel electrophoresis after in situ protease digestion on nitrocellulose. Proc Natl Acad Sci USA 84: 6970–6974

Aitkin A (1990) Identification of protein consensus sequences, Ellis Horwood, Chichester, West Sussex

Brown J, Roberts W (1976) Evidence that approximately eighty per cent of the soluble proteins from Ehrlich ascites cells are N-acetylated. J Biol Chem 251: 1009–1014

Hirano H, Komatsu S, Nakamura A, Kikuchi F, Kajiwara H, Tsunasawa S, Sakiyama F (1991) Structural homology between semidwarfism-related proteins and glutelin seed protein in rice (*Oryza sativa L.*). Theor Appl Genet 83:153–158

Hirano H, Komatsu S, Takakura H, Sakiyama F, Tsunasawa S (1992) Deblocking and microsequence analysis of N^α-acetylated proteins electroblotted onto PVDF membrane. J Biochem 111:754–757

Hirano H, Watanabe T (1990) Microsequencing of proteins electrotransferred onto immobilizing matrices from polyacrylamide gel electrophoresis: Application to an insoluble protein. Electrophoresis 11: 573–580

Hulmes JD, Pan Y-C (1991) Selective cleavage of polypeptides with trifluoroacetic acid: Applications for microsequencing. Anal Biochem 197:368–376

Ikeuchi M, Inoue Y (1988) A new photosystem II reaction center component (4.8 kDa protein) encoded by chloroplast genome. FEBS Lett 241:99–104

Krishna RG, Christopher CQ, Wold F (1991) N-terminal sequence analysis of N-acetylated proteins after unblocking with N-acylaminoacyl peptide hydrolase. Anal Biochem 199:45–50

Matsudaira P (1987) Sequence of picomole quantities of proteins electroblotted onto polyvinylidene difluoride membranes. J Biol Chem 262:10035–10038

Miyatake N, Kamo M, Satake K, Uchiyama Y, Tsugita A (1992) Removal of N-terminal formyl groups and deblocking of pyrrolidone carboxylic acid of proteins with anhydrous hydrazine vapor. Eur J Biochem 212:785–789

Moos MJr, Nguyen NY, Liu T-Y (1988) Reproducible high yield sequencing of proteins electrophoretically separated and transferred to an inert support. J Biol Chem 263: 6005–6008

Moyer M, Harper A, Payne G, Ryals J, Fowler E (1990) In situ digestion with pyroglutamate aminopeptidase for N-terminal sequencing electroblotted proteins. J Protein Chem 9:282–283

Ploug M, Stoffer B, Jensen AL (1992) In situ alkylation of cysteine residues in a hydrophobic membrane protein immobilized on polyvinylidene difluoride membranes by electroblotting prior to microsequence and amino acid analysis. Electrophoresis 13: 148–153

Tsunasawa S, Hirano H (1993) Deblocking and subsequent microsequence analysis of N-terminally blocked proteins immobilized on PVDF membrane, In: Imahori K, Sakiyama F (eds) Methods in protein sequence analysis. Plenum, New York, pp 43–51

Tsunasawa S, Sakiyama F (1992) Amino-terminal acetylation of proteins. In: Tuboi S, Taniguchi N, Katsumura N (eds) The posttranslational modification of proteins. Japan Scientific Societies, Tokyo, pp 113–121

Tsunasawa S, Takakura H, Sakiyama F (1990) Microsequence analysis of N-acetylated proteins. J Protein Chem 9:265–266

Vandekerckhove J, Bauw G, Puype M, Damme JV, Montagu MV (1985) Protein-blotting on polybrene-coated glass-fiber sheets. A basis for acid hydrolysis and gas-phase sequencing of picomole quantities of protein previously separated on sodium dodecyl sulfate/polyacrylamide gel. Eur J Biochem 152:9–19

Walsh MJ, McDougall J, Wittmann-Liebold B (1988) Extended N-terminal sequencing of proteins of archaebacterial ribosomes blotted from two-dimensional gels onto glass fiber and poly(vinylidene difluoride) membrane. Biochemistry 27:6867–6876

Wellner D, Panneerselvam C, Horecker BL (1990) Sequencing of peptides and proteins with blocked N-terminal amino acids: N-acetylserine or N-acetylthreonine. Proc Natl Acad Sci USA 87:1947–1949

Part IV
Electrophoresis and Blotting of Proteins

Chapter 13

Two-Dimensional Electrophoresis

P. JUNGBLUT

13.1
Introduction

In organelles, cells and tissues of all organisms complex functions and metabolic reactions are maintained by proteins within and at the surface of each compartment. The number of proteins occurring in a biological compartment corresponds to the complexity of the functions it has to fulfill. Ribosomes, as relatively high specialized compartments, contain in the case of *E.coli* 54 different proteins, whereas estimations for the number of expressed genes in a typical human cell are in the range of about 5000. The human genome contains about 50 000–100 000 genes. For the registration and characterization of the components of these complex systems high resolution methods are necessary. Characterization of all components of a biological compartment – the ribosomes – became a reality when methods became available for the resolution of about 50 different proteins. These methods were: first, two-dimensional electrophoresis (2-DE) (Kaltschmidt and Wittmann 1970) and, later, HPLC (Kamp and Wittmann-Liebold 1984). Today, improved 2-DE techniques play an important role in investigating the functional part of genes, i.e., proteins, within the human genome project. Furthermore changes in protein composition may be elucidated by subtractive analyses of different biological situations, for example, in disease or during differentiation.

Since the first description of protein separation using 2-DE (Smithies and Poulik 1956), the resolution of this technique has increased from 15 protein spots up to about 10|000 protein spots (Klose and Kobalz 1995). A milestone in this development was the introduction of polyacrylamide into 2-D separation by Raymond (1964). Another milestone was the combination of isoelectric focusing (IEF) with sodium dodecylsulfate polyacrylamide gel electrophoresis (SDS-PAGE) by Stegemann et al. in 1973. The enormous increase in resolution obtained by combination of

Resolution

these techniques was first presented by O'Farrell (1975), Klose (1975), and Scheele (1975). IEF and SDS-PAGE are high resolution electrophoretic methods separating up to 100 components per gel. Combination of these methods has resulted already in preliminary attempts to resolve about 1000 proteins. Attempts to perform native 2-DE with a resolution comparable to that achieved with 2-DE under denaturing conditions resulted in 2-DE patterns with low reproducibility, smearing and not very distinct protein spots. A prerequisite to achieving high resolution for cells and tissues is to reduce all S-S bridges by addition of reducing agents such as dithiothreitol (70 mM) and to cleave protein complexes completely into their polypeptide chains by preparation in 9 M urea and 3% SDS.

Major factors in obtaining improved resolution were: increased sensitivity, the use of complex ampholyte mixtures, decreased thickness of the gels (1.5–0.5 mm) and an increased size of the gels (up to 30×40 cm). In analogy to chromatography these gels are referred to as high performance two-dimensional electrophoresis (HP-2-DE)gels.

Reproducibility Whereas the outstanding resolution obtained by 2-DE may not nearly be obtained by other protein analytical methods, reproducibility is a particular problem for 2-DE separations. The reasons are: (1) the technological state of development is not comparable to that of HPLC, (2) the complexity of the separation problem is much greater than in HPLC separations, (3) the gel matrix is flexible in all directions, (4) cathodic drift effects may occur if there are any impurities in the IEF gel system, and (5) handling of the gels requires experience.

The reproducibility of protein spot silver staining intensity showed coefficients of variability in the range of 15% with a maximum value for one spot of 35%. These data were obtained for a large gel 2-DE technique using carrier ampholytes in the first dimension, comparing eight gels of one protein sample. The gels were produced by a student with 1 month of 2-DE experience. To increase reproducibility of 2-DE gels, immobilized carrier ampholytes (Immobilines) were introduced (Görg et al. 1988); however, comparable reproducibility data from Immobiline-2-DE gels (Pharmacia, Freiburg, Germany) are not available.

Reproducibility was improved by simplifications of the procedure, for example, omission of the stacking gel, the use of a continuous separation gel in the second dimension, and the use of ready-made gel solutions in IEF and SDS-PAGE (WITA, Teltow, Germany) or, in the case of Immobiline gels (Görg et al. 1988), by the introduction of ready-made gels. Initial attempts in the automatization of the whole 2-DE procedure have been made (Nokihara et al. 1992; Harrington et al. 1993).

Under optimal conditions the reproducibility of protein spot position is not a serious problem. For eight gels, the coefficients of variability of the distance between two spots were in the range of 2% with a maximum value of 3.4%. Visual and computer-assisted assignments of spots from different gels are in most cases unequivocal. Even interlaboratory comparisons are successful, as we (Jungblut et al. 1994) showed by comparing the identified spots of human heart 2-DE patterns produced in Berlin, Munich and London (Baker et al. 1992).

Proteins are detected on the gels by copper negative staining, Coomassie brilliant blue R, Coomassie brilliant blue G, silver staining or by radiochemical methods with increasing sensitivity. The detection limit for Coomassie brilliant blue R is in the range of 0.1 µg and that of silver staining in the range of 1–10 ng. The sensitivity of the detection by radioactive labeling depends on the obtained specific radioactivity of the proteins. Cellular proteins occurring with a frequency of only seven molecules per cell could be visualized on fluorograms of gels (Jungblut et al. 1987). **Sensitivity**

Proteins may be visualized after blotting on polyvinylidenedifluoride membranes or modified glass fiber membranes by Poinceau S, amido black, Coomassie blue R, or sulforhodamine staining, or with much higher sensitivity by immunostaining.

In contrast to the classical biochemical approach, which can be like looking for a needle in a haystack, subtractive analyses (Aebersold et al. 1990) allow elucidation of biological change-associated proteins, for example, disease-associated proteins, using a global approach, as shown in Fig. 1. These proteins can be analyzed by sequencing, amino acid analysis or by mass spectrometry and identified by protein database searches (Vandekerckhove et al. 1985; Aebersold 1986; Pappin et al. 1993; Zeindl et al. 1994; Knecht et al. 1994). A tumor-associated protein was identified as aldose reductase (Zeindl-Eberhart et al. 1994). Of 315 amino acids, 112 were sequenced; from these data, not only the protein name but also the right isoform, embryonic aldose reductase, could be assigned to the variant protein. In dilated cardiomyopathy, four proteins were found to be disease-associated, with changes in spot intensity seen in human myocardium 2-DE gels: creatine kinase, malate dehydrogenase, α-crystallin B, and heat-shock protein 27 (Knecht et al. 1994; Otto et al. 1994). **Subtractive analysis**

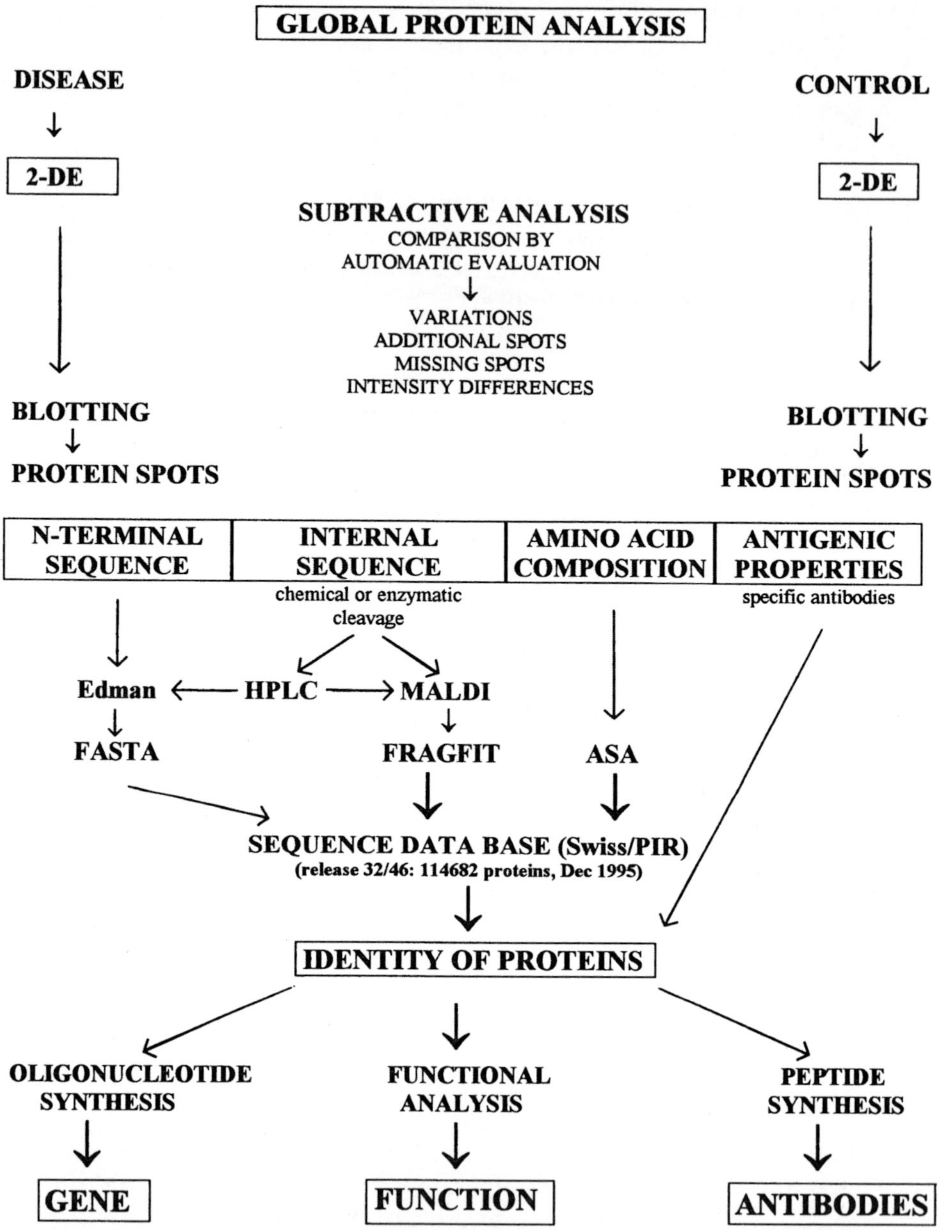

GLOBAL PROTEIN ANALYSIS

DISEASE   CONTROL
↓   ↓
2-DE   2-DE

SUBTRACTIVE ANALYSIS
COMPARISON BY
AUTOMATIC EVALUATION
↓
VARIATIONS
ADDITIONAL SPOTS
MISSING SPOTS
INTENSITY DIFFERENCES

BLOTTING   BLOTTING
↓   ↓
PROTEIN SPOTS   PROTEIN SPOTS

N-TERMINAL SEQUENCE	INTERNAL SEQUENCE	AMINO ACID COMPOSITION	ANTIGENIC PROPERTIES

chemical or enzymatic   specific antibodies
cleavage

Edman ⟵ — HPLC ⟶ MALDI

FASTA   FRAGFIT   ASA

SEQUENCE DATA BASE (Swiss/PIR)
(release 32/46: 114682 proteins, Dec 1995)

IDENTITY OF PROTEINS

OLIGONUCLEOTIDE   FUNCTIONAL   PEPTIDE
SYNTHESIS   ANALYSIS   SYNTHESIS

GENE   FUNCTION   ANTIBODIES

Fig. 1. Global protein analysis. *PIR*, Protein Information Resource; *FASTA*, search program for sequences in the sequence database; *FRAGFIT*, search program for mass data obtained by mass spectrometry in the sequence database; *ASA*, search program for amino acid composition data in the sequence database

The evaluation of 2-DE gels may be carried out visually or auto- **Evaluation** matically with the help of a computer. Visual evaluation is very **of 2-DE gels** precise, with differences in protein spot staining intensity of 10% detected reliably (Jungblut et al. 1986). Furthermore, visually finding corresponding spots in different gels is much easier than with automated matching. Visual evaluation is not dependent on expensive hard- and software, but it becomes increasingly laborious the more protein spots there are to be investigated and the larger the gels are. For the comparison of 20 gels with sizes of 30×40 cm, it is nearly impossible to compare the intensity of a single spot in the whole data set, whereas by automatic evaluation the interesting spot may be zoomed out and compared on 20 windows on the monitor, at the same time allowing control of the calculated intensity differences. Maximum gel quality, optimized spot detection parameters and the possibility of interactive changes are necessary to use automatic evaluation. The process may be divided into five steps: (1) digitalization of the data after scanning, (2) background and smear corrections, (3) spot detection, (4) quantification and (5) comparison of different gels (matching). Different automatic evaluation systems are commercially obtainable. It should be noted that the sensitivity of spot detection cannot be increased by changing from visual to automatic evaluation without inclusion of artifacts (Prehm et al. 1987).

Criteria for the efficiency of an automated evaluation system are: local resolution, dynamic range of optical density, time for each software procedure, quality of spot detection and matching procedure, possibilities of interactive operations, and price.

The immense amount of data obtained by subtractive analyses, **2-DE databases** identification and characterization of proteins on 2-DE gels are stored in protein 2-DE databases, published once a year in *Electrophoresis*; also, some of them are accessible in the Internet using World Wide Web technology. For example a 2-DE database of human heart proteins was built up containing 3239 proteins from which, at this writing, 66 have been annotated (Jungblut et al. 1994). Annotation criteria used in this database are dilated cardiomyopathy-associated proteins, NH_2-terminal sequence, internal sequence, amino acid composition, protein name, molecular mass, isoelectric point, and reference for identification. Despite the fact that the myocardial proteins of Baker et al. 1992 were prepared by a different method and that both another 2-DE gel system and another gel size had been used, 40 of the proteins characterized in this database could be easily transferred to our database.

Gel size As mentioned above the gel size is a major factor in the resolution of 2-DE gels. Whereas with small gels (6.5 × 7.5 cm) resolutions of up to 1000 protein spots can be obtained, on large gels (30 × 40 cm) up to 10|000 proteins can be separated. Here we describe a small gel technique (Jungblut and Seifert 1990), which is easy to establish and which is sufficient for complex protein mixtures of up to 1000 protein species. Much information on protein composition changes can be obtained by subtractive analyses at the organelle level of complexity. Clearly, compared to sera testing by one-dimensional tests, more information can be derived from blots of microorganisms using this simple 2-DE technique. Furthermore purity tests of purified proteins can be done. The use of ready-made gel solutions reduces the time-consuming search for mistakes in case of poor resolution on the gels. The recipes can also be used for larger gels, but the expense increases logarithmically with increasing size.

13.2
Flow Chart

– First day:

  Sample preparation

  Preparation of gel rods for IEF

  Second day:

  Sample application

  IEF run

  Gel preparation SDS-PAGE

  IEF gel equilibration

  Sample loading

  Run SDS-PAGE

  Fixation (overnight)

  Third day:

  Staining

Samples can also be prepared and stored frozen at –70 °C or in liquid nitrogen. The schedule shown is for silver staining. For Coomassie blue R-250 staining, fixation can be reduced to 1 h, with staining overnight and destaining on the third day. If the SDS-PAGE gels are prepared for blotting, the procedure can be interrupted after IEF gel equilibration. Equilibrated IEF gels can be stored frozen at –70°C for a few months. On the third day SDS-PAGE, blotting and staining is performed. Interrupting after SDS-PAGE and before blotting is not recommended.

13.3
Sample Preparation for 2-DE

13.3.1
Solid Probes

9 M urea	108 mg	urea	**Lysis buffer**
70 mM DTT	+ 10 µl	1.4 M DTT	
2% ampholyte (Servalyte 2–4)	+ 10 µl	40% Ampholyte Serva, (Servalyte 2–4)	
	100 µl	deionized water	
	200 µl	lysis buffer	

Lysis buffer is added to probes in the required volume/mass ratio. For single proteins the concentration should be 0.5–1 mg/ml; for complex mixtures of proteins the desired concentration is 10–30 mg/ml. **Procedure**

The sample is kept for 30 min at room temperature in lysis buffer and from time to time gently stirred with the pipet tip used for adding the lysis buffer. After solubilization the sample is frozen and stored at –70 °C.

13.3.2
Liquid Probes

First, the volume of probe (Y) in microliters is determined. The numerical value of this volume is denoted as factor Y. Using this factor, the addition of urea, DTT and ampholytes is calculated:

$Y \times 1.08 =$ mg urea to be added
$Y \times 0.1 =$ µl 1.4 M DTT to be added
$Y \times 0.1 =$ µl 40% ampholyte (Servalyte 2–4) to be added

For example: for 100 µl probe 108 mg urea, 10 µl 1.4 M DTT and 10 µl 40% ampholyte (Servalyte 2–4) have to be added.

After this adjustment the original probe volume is doubled and protein concentration is decreased to one half.

13.3.3
Tissue Homogenization

AP buffer

AP buffer (derived from the German *Aufarbeitungspuffer*) has the following composition: 9 M urea, 2% ampholyte, ph 2–4 (Servalyte), 70 mM DTT, 25 mM Tris/HCl (pH 7.1), 50 mM KCl, 3 mM EDTA, 2.9 mM benzamidine, 2.1 μM leupeptin. Better protection against the effects of protease is achieved by adding 0.02 volumes of 5 μM pepstatin, 50 mM PMSF in ethanol to the sample (after prior addition of AP buffer). Both AP buffer and pepstatin/PMSF solution are supplied by WITA (Teltow, Germany).

Tissue pieces < 20 mg

Tissue pieces < 20 mg are placed frozen in Eppendorf tubes. The tubes are cooled by liquid nitrogen, while the probes are homogenized using a pointed glass pestle (WITA, Teltow, Germany) that fits exactly into the tubes. During thawing lysis buffer or AP buffer is added in the required volume/mass ratio. Buffer volume/tissue mass ratios higher than 6:1 are recommended to preserve the 9 M concentration of urea in the sample.

Tissue pieces > 20 mg

Tissue pieces > 20 mg are placed into an agate or quartz mortar (WITA, Teltow, Germany). The mortar is chilled in liquid nitrogen and the sample is homogenized with a cold pestle (WITA, Teltow, Germany). After thawing, lysis buffer or AP buffer is added in the required volume/mass ratio (> 6:1). The replacement of lysis buffer by AP buffer is recommended in order to preserve ionic strength and to prevent proteolytic degradation of the sample.

13.4
Solutions for 2-DE

First dimension: IEF

1. Prepare 0.8% APS: weigh out 0.8 g of ammonium persulfate (APS) and bring to 100 ml with deionized water. Divide into 60 μl aliquots, store frozen at −70 °C.

2. IEF separation gel: 4% (w/v) acrylamide, 0.3% (w/v) piperazine diacrylamide, 2% WITAlyte (pH 2–11), 9.0 M urea, 5% (w/v) glycerol, 0.06% (v/v) TEMED, 0.02% (w/v) APS. Shortly before application, mix 1365 μl degassed separation gel solution (WITA, Teltow, Germany) with 35 μl of 0.8% APS.

3. IEF-cap gel: 10% (w/v) acrylamide, 0.13% (w/v) piperazine diacrylamide, 2% WITAlyte (pH 2–11), 9.0 M urea, 5% (w/v)

glycerol, 0.06% (v/v) TEMED, 0.02% (w/v) APS. Shortly before application mix 390 μl degassed cap gel solution (WITA, Teltow, Germany) with 10 μl of 0.8% APS.

4. Prepare anode solution: 7.27% (w/v) phosphoric acid, 3.0 M urea.

   Note: The following ingredients should be freshly prepared. Weigh out 45 g urea and place in a beaker. Fill graduated cylinder with deionized water to 237.5 ml; pour water into the beaker and solubilize by using a magnetic stirrer.
   Seal the beaker with Parafilm. Immediately before using add 12.5 ml of 85% (m/v) phosphoric acid and stir by magnetic stirrer.

5. Cathode solution: 5% (v/v) ethylene diamine, 9.0 M urea, 5% (w/v) glycerol.

   Note: The following ingredients should be freshly prepared. Weigh out 135 g urea and 12.5 g glycerol. Bring to 237.5 ml with deionized water.
   Stir using a magnetic stirrer in the beaker, heating to 40 °C until urea and glycerol are solubilized. Add 12.5 ml ethylendiamine immediately before use; stir by magnetic stirrer.

6. Prepare overlaying solution: 5% (w/v) glycerol, 5 M urea, 2% (w/v) Servalyte 2-4. Weigh out 6 g urea and 1 g glycerol and bring to 12 ml with deionized water. Remove 3.8 ml from this stock solution and mix them with 0.2 ml Servalyte 2-4. Divide into 200 μl aliquots.

7. Prepare Sephadex gel solution (WITA, Teltow, Germany) containing 70 mM DTT.
   Prior to the application, mix 100 mg Sephadex gel solution with 108 mg urea and 10.0 μl of Servalyte 2-4 mixture. Shake thoroughly for 10–15 min using a Vortex.

8. Prepare equilibration solution: 125 mM Tris/phosphate (pH 6.8), 40% (m/v) glycerol, 65 mM DTT, 3% (m/v) SDS.

8a. To make buffer: weigh out 12.11 g Tris-base and bring to 35 ml with deionized water. Adjust pH to 6.8 using 1 M phosphoric acid and bring to 100 ml with deionized water.

8b. To make solution: Add 160 g glycerol and 12 g SDS to 50 ml of buffer. Bring to 400 ml with deionized water, freeze at –70 °C in 20 ml aliquots.
   Directly before use add to each aliquot 0.2 g DTT and solubilize.

9. Prepare 1.28% (m/v) APS: Weigh out 1.28 g ammonium persulfate and bring to 100 ml with deionized water. Divide into 1.2 ml aliquots.

10. Prepare agarose: 1% (w/v) agarose (low melt preparative grade, BioRad, Munich, Germany), 125 mM Tris/phosphate (pH 6.8), 0.1% (w/v) SDS. Measure out 1.25 ml of equilibration buffer (see step 8). Add 8.7 ml of 0.115% (m/v) SDS solution and 0.1 g agarose. Solubilize under 70 °C. For use immediately before the application store at 40 °C, for longer times store at 4 °C.

11. Prepare SDS-PAGE gel: 15% (w/v) acrylamide, 0.2% (w/v) bis-acrylamide, 375 mM Tris-HCl (pH 8.8, 20 °C), 0.03% (v/v) TEMED, 0.1% (w/v) SDS, 0.08% (w/v) APS. Before the application add 1.2 ml of 1.28% APS to 18 ml of SDS-PAGE gel solution (WITA, Teltow, Germany).

12. Electrode buffer for second dimension: 25 mM Tris, 192 mM glycine, 0.1% (w/v) SDS.

    Note: Use freshly prepared ingredients. Weigh out 2.422 g Tris-base, 11.52 g glycine and 0.8 g SDS. Bring to 800 ml with deionized water and solubilize in a graduated cylinder using a magnetic stirrer.

13. Prepare fixation solution: 50% (v/v) ethanol or methanol, 10% (v/v) acetic acid. Combine 500 ml methanol or ethanol with 100 ml acetic acid and 400 ml deionized water.

13.5
First Dimension (Isoelectric focusing, IEF)

13.5.1
Preparation of Gel Rods

Materials   The following chemicals and materials for one run (eight rods) should be prepared in advance:

- 2×gel casting apparatus (WITA, Teltow, Germany)

- 1365 µl separation gel solution (WITA, Teltow, Germany)

- 390 µl cap gel solution (WITA, Teltow, Germany)

- 0.8.% ammonium persulfate solution (APS) (35 µl for separation gel, 10 µl for cap gel)

- deionized water

- Pasteur capillaries

- Gilson pipettes: 0–20 µl, 0–200 µl, gel loader tips

- eight propylene (PP) threads with 1.4 mm diameter with sealed edges (WITA, Teltow, Germany)

- gel tubes for gels: 1.5 mm inner diameter, length 9.3 cm (WITA, Teltow, Germany)

- calibration rings: 4 mm and 10 mm from the bottom, 13, 17 and 23 mm from the top

- pump for degassing of gel solution

- lab clocks

- Parafilm: cut into small squares

- strips of filter paper (width < 1.5 mm)

- ruler

1. Insert four glass tubes, with cap gel end (2 calibration rings: 4 mm and 10 mm) below, into the holders of each of the two gel casting apparatuses such that the glass tubes stand just above the "filling boat" within one of the two compartments. Pass PP threads from the upper end of the glass tubes to the lower end (careful, the threads are not easily moveable); otherwise it will not be possible later on to use them for drawing the gel solutions up. **Procedure**

2. Thaw the separation gel solution and ammonium persulfate solution. The temperature of the gel solution has to be monitored, since at high temperatures (> 25 °C) polymerization may be too fast.

3. Degas the separation gel solution for 4 min; the glass tubes are gently tapped on the edge of the table to prevent air bubbles.

4. Thaw the cap gel solution and degas as in step 3.

5. Using a Gilson pipette add 35 µl APS to 1365 µl separation gel solution. The solution should be carefully shaken to avoid introduction of oxygen into the gel.

6. Fill separation gel solution into the filling boat (700 µl per 4 tubes) of each of the two gel casting apparatuses. The ends of all the glass tubes must be immersed in the gel solution.

7. Draw gel solution up to the 23 mm mark by carefully pulling out the PP threads.

8. Change the filling boat compartment.

9. Using a Gilson pipette, add 10 µl 0.8% APS to 390 µl cap gel solution. Mix by gentle shaking.

10. Pour 200 µl cap gel solution into the free compartments of the filling boats of each of the two gel casting apparatuses; the gel solution should reach as far as the 17 mm mark.

11. Remove filling boat and allow gel solutions to extend to the 13 mm mark. By this step air is sucked into the tube between the lower end and the 4 mm mark.

12. After 30 min polymerization, pull out PP threads and remove non-polymerized solution at the upper surface of the gel. Form a moist chamber on the upper side by adding a large drop of deionized water on the aperture of the capillary. An air bubble is formed between the gel and the drop of water. The moist chamber formed in this way and the cap gel side are subsequently closed by Parafilm. For further polymerization the gels should be kept at room temperature overnight. The prepared tubes can be used the day after or as late as 4 days after their preparation if they are stored at room temperature.

13.5.2
Loading the Sample

Materials The chemicals and materials should be prepared in advance. If starting in the morning with sample application, the urea for the anode and cathode solutions should be dissolved the day before. Phosphoric acid and ethylenediamine should be added directly before starting sample application.

- prepared gel tubes
- IEF chamber (WITA, Teltow, Germany)
- anode solution
- cathode solution
- prepared sample
- Sephadex solution (WITA, Teltow, Germany)
- urea
- mixture of ampholytes (WITAlytes 2–11)
- overlaying solution
- deionized water

- Pasteur pipettes

- 0–20 µl Gilson pipettes

- 0–200 µl Gilson pipettes

- gel loader pipette tips

- balance

- vortex

- filter paper strips (width < 1.5 mm)

1. Ethylenediamine is added to the cathode solution and de- **Procedure**
   gassed for 5 min. The bottom chamber is filled with cathode
   solution.

2. Thaw the Sephadex solution. Add 108 mg urea and 10 µl of
   the ampholyte mixture to 100 mg Sephadex gel. Shake the
   mixture thoroughly by vortexing for 10–15 min.

3. Remove the tubes from the gel casting apparatus and insert
   them into the anode part of the focusing chamber. The 13 mm
   designation should be still visible. Cap gel side faces upwards.

4. Fill cathode solution into the cathode side of the tubes without bubbles (after prior removal of water).

5. Mount the anode part of the focusing chamber onto the bottom part of the focusing chamber. The tubes must dip into the
   cathode solution.

6. Remove water and dry the loading side as well. Water is removed using an extended Pasteur pipette or a gel loader tip
   and the gel surface is subsequently dried with filter paper.

7. Thaw the prepared sample and overlaying solution.

8. Using gel loader pipette tips, quickly overlay the gel with approximately 2 mm prepared Sephadex (see step 2).

9. Using a gel loader tip, load the sample (1–15 µl), applying it
   onto the Sephadex surface. The tip must contact the Sephadex
   layer at the moment the sample is pressed out. The sample
   should be loaded carefully, without air bubbles.

10. Gently layer 5 µl of overlay solution onto the sample using gel
    loader tips; avoid mixing of overlay solution and sample. The
    capillary is subsequently filled with anode solution (freshly
    prepared by addition of phosphoric acid; see solutions for 2-
    DE) without air bubbles.

11. The remaining anode solution is used to fill the upper chamber, covering all tubes with buffer.

13.5.3 IEF Run

Materials   – power supply (voltage range up to 1000 V)

– protocol sheets

– clock (if no programable power supply is available)

Procedure Voltage is increased in stepwise manner to obtain a Vh product of 1841 according to the following table:

Voltage	Time
100 V	75 min
200 V	75 min
400 V	75 min
600 V	75 min
800 V	10 min
1000 V	5 min

13.5.4 Gel Equilibration

Materials   – equilibration solution

– 87% glycerol

– deionized water

– small Petri dishes, diameter 5 cm

– 5 ml glass pipette with Peleus ball

– polypropylene (PP) threads, diameter 1.5 mm (WITA, Teltow, Germany)

Procedure 1. Thaw the equilibration solution (about 1 h before the end of the IEF run). Before using, add 0.2 g DTT to 20 ml of equilibration solution.

2. After the end of the IEF run, remove the cathode and the anode solutions from the tubes.

3. Overlay the cap gel side with 87% glycerol solution.

4. Fill 5 ml of equilibration solution into the Petri dishes.

5. Extrude the gels using PP threads: PP threads are inserted into the tubes from cap gel side. While the loading side of the tube is dipped into deionized water, the gel is carefully expelled

slightly downwards. At the same time the dissolving of over-lay solution in water can be observed. After the gel has been pulled out about 5 mm, the extruded end is briefly washed with water. The tube is then held above the Petri plate with equilibration solution and the whole gel is completely ex-truded by the PP thread into the plate. Two gels may be equilibrated within one plate.

6. Equilibrate the gel **exactly 10 min** with shaking at room tem-perature.

7. If the gels are not used immediately after IEF, the equilibra-tion solutions are poured off and the gels are stored in Petri plates at –70 °C. Storage at –20 °C results in lower resolution; storage in liquid nitrogen destroys the gels.

13.6
Second Dimension (SDS-PAGE)

13.6.1
Gel Preparation SDS-PAGE

Materials

The materials, solutions and chemicals must be prepared in ad-vance. Amounts are described for the preparation of two small gels. Agarose must be solubilized at 70 °C and kept solubilized at 40 °C until it is used for embedding the IEF gels. Using four SDS-PAGE chambers, all of the eight IEF gels may be run in parallel.

- 70% (v/v) ethanol
- deionized water
- Kimwipes
- Mini Protean II chamber (BioRad, Munich, Germany), includ-ing electrode assembly and lid with cables, glass plates (two small plates and two large ones), four spacers (1.5 mm), pair of clamp assemblies, casting stand with rubber gasket at the bottom end
- ruler
- Pasteur pipettes
- agarose solution
- 18 ml of SDS-PAGE gel solution (WITA, Teltow, Germany)
- 1.2 ml 1.28% APS
- clock

– water bath with thermometer

Procedure   1. Heat agarose solution up to 70 °C to liquify it.

2. Cool the agarose solution to 40 °C to keep it liquid.

3. Clean glass plates and spacers by sprinkling with 70% ethanol and wiping cleaned with Kimwipes.

4. Thaw the gel solution and the APS.

5. Lock the glass plates into the clamp assembly, which is placed in the front slot of the casting stand so that the screws face the casting stand and "horns" face upwards. Turn the screws far enough so that the thick plexiglass plates are aligned precisely along the rear margin. The larger glass plate can now be put on the front side of the plexiglass plate. Then the spacers are placed along the right and the left margins. Insert the small plate. With a gentle pressure of the thumb on the plates and spacers, make sure that the glass plates lie on the bottom. Tighten the formed sandwich with the clamp screws. The presence of Newton's rings on the margin of the rear side means that the screws are quite firmly tightened. Do not tighten the screws too firmly, otherwise the glass plates will break. Pull the sandwich from the casting stand and determine by thumb if the glass plates and spacers are flush at the bottom.

6. Mark 6.9 cm from the bottom of the small glass plate as filling aid (corresponds to a gel length of 6.6 cm).

7. Lock the assembled gel sandwich onto the casting stand. The screws face inward, "horns" upward. By pressure on the plexiglass plate the gel sandwich is inserted under the small protrusion.

8. Mix 18 ml of gel solution with 1.2 ml APS without foaming. The solution must be homogeneous; avoid the introduction of oxygen.

9. The vessel containing the gel solution is placed at the rear glass plate above the filling groove so that the gel solution can flow down the glass along the groove without forming air bubbles. The separation gel is poured up to the marking line.

10. The gel solution is immediately overlaid with deionized water to support the polymerization process and to obtain a smooth surface. Use a Pasteur pipette and work at one end of the gel to prevent swirling of gel solution and to stratify water ho-

mogenously along the whole gel surface. After 60 min the gels are ready for running SDS-PAGE.

13.6.2
Sample Loading

The following solutions and materials must be prepared in advance: **Materials**

– agarose solution (40 °C)

– electrode buffer (800 ml)

– gels of 1st dimension (IEF gels)

– Mini Protean II chamber (BioRad, Munich, Germany)

– filter paper

– small spatula

1. Remove deionized water from gel surface using filter paper.   **Procedure**

2. Insert assembled gel sandwiches into electrode assembly (screws face out, horns upwards): first insert the assembled gel sandwiches into the upper part of the electrode assembly. The bottom part is then fixed by pressing it against the electrode assembly.

3. Load the IEF gels on the surface of the slab gels: Remove the rod from the equilibration buffer and lay it lengthwise along the margin of the thick plexiglass plate. Try to place the rod gels as close as possible to the edge of the filling groove. Using a spatula, gently push the rod gel down so that it slides carefully into the filling groove. The rod gel must lie on the surface of the SDS-PAGE gel without air bubbles.

4. Using a Pasteur pipette, fill the filling groove with 40 °C heated agarose (without air bubbles) as far as the upper edge. The agarose should be solidified after 2 min.

5. After installation of the electrode assembly with assembled gel sandwiches, the BioRad chamber is filled with 620 ml of electrode buffer. Avoid the formation of foam and air bubbles under the electrode assembly during filling. The level of electrode buffer in the inner chamber should cover the red plastic spot on the electrode assembly.

13.6.3
SDS-PAGE Run

Materials – power supply (150 V; 100 mA, if 4 chambers are run in parallel at least 300 mA are necessary)

– Mini Protean II chamber (BioRad, Munich, Germany)

– protocol sheet, pencil

– clock (if power supply is not programmable)

– gloves

– fixation solutions: 50% methanol/10% acetic acid/40% water or 50% ethanol/10% acetic acid/40% water

Procedure 1. Before closing the chamber, it is necessary to determine if the surface of the plexiglass is dry to prevent formation of "crawling" currents. The gel surface must be covered by electrode buffer.

2. The voltage is increased in a stepwise manner according to the following table; the current and power values are set to the maximum of the power supply. The current values are then as indicated in the table and are decreased within each step. The beginning and the end values within a given time interval are given in the table:

Voltage	Time	Current one chamber (2 gels)	Current two chambers in parallel (4 gels)
35 V	5 min	22–21 mA	44–42 mA
55 V	10 min	33–32 mA	66–64 mA
100 V	15 min	59–48 mA	118–96 mA
150 V	60 min	72–50 mA	144–100 mA

3. After the end of SDS-PAGE, pour off the electrode buffer and remove the assembled gel sandwiches from the electrode assembly. To remove the gel, gently twist the spacer so that the small glass plate pulls away. Now the gels can be transferred to fixation solution.

   Note: Use gloves! Before using them wash the gloves with water to remove talcum powder.

4. After fixation various staining techniques can be applied.

13.7
Staining with Coomassie Brilliant-Blue R-250

The following Coomassie blue staining procedure was described by Eckerskorn et al. (1988). For 1.5 mm thick gels, 100 µg of a complex protein mixture from different mammalian organs has been enough to obtain complex 2-DE patterns. For Coomassie brilliant-blue R-250, a minimum of about 100 ng/protein species is necessary for detection.

- staining dishes   **Materials**

- gloves

- Coomassie blue R-250

- ethanol: denatured by 1% methylethylketone

- methanol: Merck 6009

- acetic acid

The following amounts are recommended for staining and   **Solutions**
destaining of four 6.5 × 7.5 cm gels.

- fixation solution: 50% ethanol/10% acetic acid/40% deionized water (250 ml ethanol + 50 ml acetic acid + 200 ml deionized water)

- staining solution: 50% methanol/10% acetic acid/0.05% Coomassie blue R-250 (250 ml methanol + 50 ml acetic acid + 200 ml deionized water + 0.25 g Coomassie blue R-250; with new Serva blue R-250 only 70% of given amount is needed)

- destaining solution: 5% methanol/12.5% acetic acid/82.5% deionized water (25 ml methanol + 62.5 ml acetic acid + 412.5 ml deionized water)

- preserving solution: 7% acetic acid (140 ml acetic acid + 1860 ml deionized water)

   The destaining and preserving solutions can be regenerated by charcoal filtration.

Most of each protein is located in the inner part of the gel. For   **Procedure**
quantitative staining it is therefore important that the dye reaches the inner part of the gel. For 1.5 mm thick gels with 15% acrylamide, staining times of at least 6 h are recommended. Staining for 1 h may be enough for qualitative detection of protein amounts over 500 ng. For two 6.5×7.5 cm gels, 250 ml of each solution are sufficient.

1. Gels are gently shaken overnight or at least 1 h in fixation solution

2. Gels are stained by gentle shaking in staining solution overnight or at least 1 h.

3. To remove the background, the gels are destained in destaining solution by gentle shaking for at least 1 h.

4. Further destaining is carried out in preserving solution for at least 4 h. The solution should be changed twice during the first hour and once each hour thereafter.

   Note: The above four steps of this type of staining are performed at room temperature. The staining should be performed with methanol because the Serva blue concentration is based on this type of solvent.

5. Preserving the sealed gels at 4 °C in preserving solution is possible for at least 5 years.

13.8
Silver Staining

Detection of proteins in polyacrylamide gels by silver staining is the most sensitive unspecific staining method. It was first described by Merril et al. in 1979, with a general overview provided by Rabilloud et al. in 1994. Three different types of silver staining, diamine or ammoniacal stains, nondiamine chemical reduction stains and photodevelopment stains, are used. Nondiamine chemical reduction staining was optimzed by Heukeshoven and Dernick (1985). This method is the basis for the procedure described here. It was optimized for the 1.5 mm thick 2-DE gels used here (Jungblut and Seifert 1990). This method is more sensitive than the photodevelopmental stains and as sensitive but simpler than the diamine silver stain. Protein spots containing 1–10 ng of protein are detectable on the 2-DE gels used here. Loading of a minimum of 15 μg of a complex mixture of proteins, e.g., from human heart, is sufficient to detect some 100 proteins on 6.5 × 7.5 cm 2-DE gels.

Materials
- staining dishes (opaque white or transparent white plastic or glass dishes)

- gloves

- EDTA (Titriplex III; Merck, Darmstadt, Germany)

- ethanol (Merck, Darmstadt, Germany)

- formaldehyde (35%, Merck, Darmstadt, Germany)
- glutaraldehyde for electron microscopy (25%, Merck, Darmstadt, Germany)
- acetic acid (Merck, Darmstadt, Germany)
- sodium acetate (Merck, Darmstadt, Germany)
- anhydrous sodium carbonate (Merck, Darmstadt, Germany)
- anhydrous sodium hydrogen carbonate (Merck, Darmstadt, Germany)
- sodium thiosulfate (Merck, Darmstadt, Germany)
- silver nitrate (Merck, Darmstadt, Germany)
- Thimerosal (Sigma, St. Louis, USA)

Amounts are given for four 6.5 × 7.5 cm gels. All of the solutions, **Solutions** except the fixation solution, are prepared just prior to use. The use of fresh glutaraldehyde as specified above is essential for successful staining. Glutaraldehyde is stored in 20 ml aliquots at 4 °C.

- fixation solution: 50% ethanol/10% acetic acid/40% deionized water (1000 ml ethanol + 200 ml acetic acid + 800 ml deionized water)

- incubation solution: 30% ethanol/0.5 M sodium acetate/0.5% glutaraldehyde/0.2% sodium thiosulfate (300 ml ethanol + 68.04 g sodium acetate + 20 ml 25% glutaraldehyde + 2 g sodium thiosulfate. Bring to 1000 ml with deionized water)

- silver nitrate solution: 0.1% $AgNO_3$/0.01% formaldehyde

  Solubilize 1 g $AgNO_3$ in 1000 ml of deionized water and add 288 µl of 35% formaldehyde.

- developer: 2.5% sodium carbonate (pH 11.3)/0.05% sodium thiosulfate/0.01% formaldehyde

  Combine 25 g sodium carbonate and 12 mg sodium thiosulfate and bring to a final volume of 1000 ml with deionized water. Add 288 µl of 35% formaldehyde and adjust pH to 11.3 using sodium hydrogen carbonate.

- stop solution: 0.05 M EDTA (Titriplex III)/0.02% Thimerosal (18.612 g EDTA fill up to 1000 ml with deionized water + 200 mg Thimerosal)

Procedure   All steps should be performed at a constant room temperature, especially the developing step. For two 6.5 × 7.5 cm gels, 250 ml of each of the following solutions are sufficient.

1. Shake gels in fixation solution overnight or at least 1 h.

2. Gently shake gels in incubation solution 2 h.

3. Wash gels in deionized water 3x, 20 min each.

4. Gently shake gels in silver nitrate solution for 30 min.

5. Briefly (30 s) wash gels with deionized water.

6. Develop gels by shaking 5–30 min in developer. The time is dependent on the amount of loaded proteins.

7. If required color intensity is reached, the gels are transferred to stop solution for at least 15 min.

8. Exchange the stop solution. The gels can be stored in the dark at 4 °C in stop solution for several months.

13.9
Cleaning of IEF Tubes

The tubes used for isoelectric focusing must have an exact inner diameter. Otherwise casting and removing of the gels using the PP threads does not work. For this mechanical reason but also to avoid cathodic drift effects, the inner surface has to be free of any impurities. Therefore, the following cleaning procedure is strongly recommended.

Materials   – 500 ml glass beaker

– 100 ml glass beaker

– stable wire from inert material

– Deconex 12 PA (Borer Chemie, Zuchwil, Switzerland)

– 0.1 M HCl

– deionized water

– tweezers

– heated plate

Procedure   1. After the end of IEF the tubes are thoroughly rinsed out with deionized water.

2. Heat 500 ml of a 6% Deconex solution to 60°–80 °C.

3. Dip the glass tubes in the Deconex solution and leave at 70 °C for 3 × 10 min: After 10 min, individually remove the glass tubes using tweezers. Allow the cleaning solution to drain out and place the tubes in fresh cleaning solution again for 10 min at 70 °C.

4. Rinse the tubes thoroughly with deionized water.

5. Heat 500 ml of 0.1 M HCl solution to 95 °C.

6. Place the tubes in a glass beaker and add 0.1 M HCl until the beaker is full and tubes are covered. Leave the tubes in HCl solution for 30 min at 95 °C.

7. Rinse the tubes with deionized water and allow to dry.

13.10
Results

The 2-DE technique described here has been used for a wide range of probes from prokaryotes to eukaryotes. Two examples are shown in Figs. 2 and 3. Using 2-DE gels, development of the fungus *Uromyces viciae-fabae* from spores to the haustorium mother cell has been investigated by comparing the protein composition during these six stages of differentiation (Deising et al. 1991). Some of the protein spots disappeared, some were newly synthesized, others were expressed in higher or lower amounts. Mouse brain proteins were investigated to elucidate genetic variants between two inbred lines. To increase resolution, biological and biochemical prefractionation was used before 2-DE. Qualitative genetic variability in mitochondria (Jungblut and Klose 1985), nuclei (Jungblut et al. 1989), cytosol, membranes, heparin-binding (Jungblut and Klose 1986), dye-binding and metal-binding proteins was found for about 1% of the analyzed proteins, whereas quantitative genetic variability was in the range of 10%. Five of the mouse brain proteins identified (Eckerskorn et al. 1988; Jungblut et al. 1992) are shown in Fig. 3. A silver stained 2-DE pattern of human myocardium is shown in Fig. 4. The pattern contains about 500 spots, from which about 100 have been identified (Jungblut et al. 1994). Some of the identified proteins are indicated; α-crystallin B, malate dehydrogenase, creatine kinase and heat-shock protein 27 have been shown to be associated with dilated cardiomyopathy and varied in amount (Knecht et al. 1994; Otto et al. 1994).

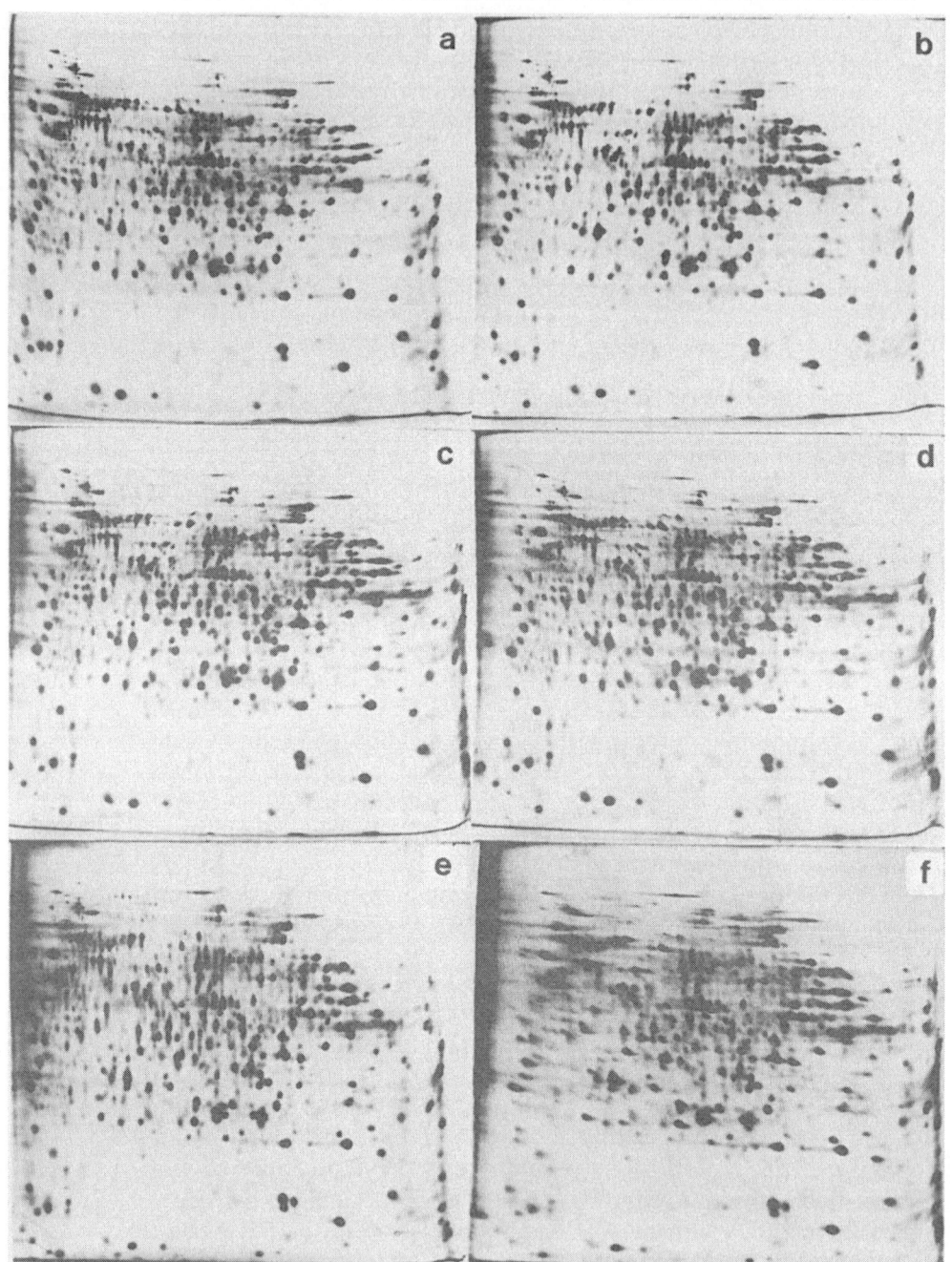

Fig. 2a–f. Silver-stained 2-DE patterns of *Uromyces viciae-fabae* uredospores (a), germ tubes (b) and infection structures differentiated for 7 h (c), 9 h (d), 18 h (e) and 24 h (f)

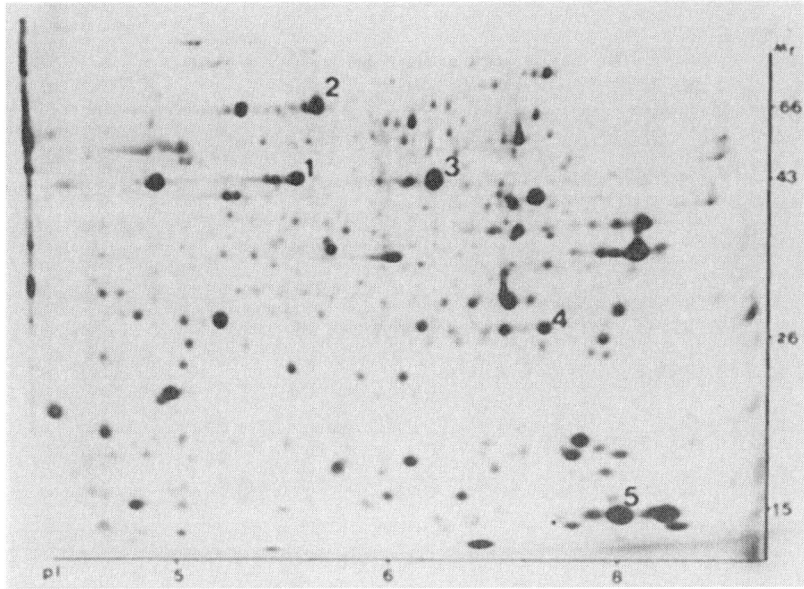

Fig. 3. Coomassie brilliant blue R-250 stained 2-DE pattern of mouse brain proteins. 140 µg of protein were applied to IEF. Calibration of pI and Mr was performed using internal marker proteins. Five of the proteins identified by NH₂-terminal sequencing and/or amino acid analysis are indicated with numbers. *1*, actin; *2*, serum albumin; *3*, creatine kinase; *4*, triose phosphate isomerase; *5* β-hemoglobin

Using the 2-DE technique described here, complex patterns have been obtained for all of the probes analyzed so far with one exception: ribosomal proteins of *E.coli*. These proteins are extremely basic and therefore the pH gradient is not optimal. The slope of the ampholyte gradient is low in the pH 5–7 range and high between pH 3.5–5 and pH 7–10. Nevertheless, ribosomal proteins can be resolved by shortening the Vh product to about one fourth.

Good reproducibility is a prerequisite for subtractive analyses and readily obtained with the 2-DE technique. The mouse brain patterns shown in Fig. 5 were reproducibly obtained by a student 12 times after 2 months of intensive training.

Calibration of pI values is obtained by measuring pH values with a surface electrode or by cutting off pieces of the gel and measuring the pH value of a water suspension obtained after sonication (Klose and Kobalz 1995). Calibration is also obtained by the use of marker proteins. Often, native pI values or those calculated from the sequence are used, neglecting the fact that the pI value changes in 9 M urea solutions. For each calibration,

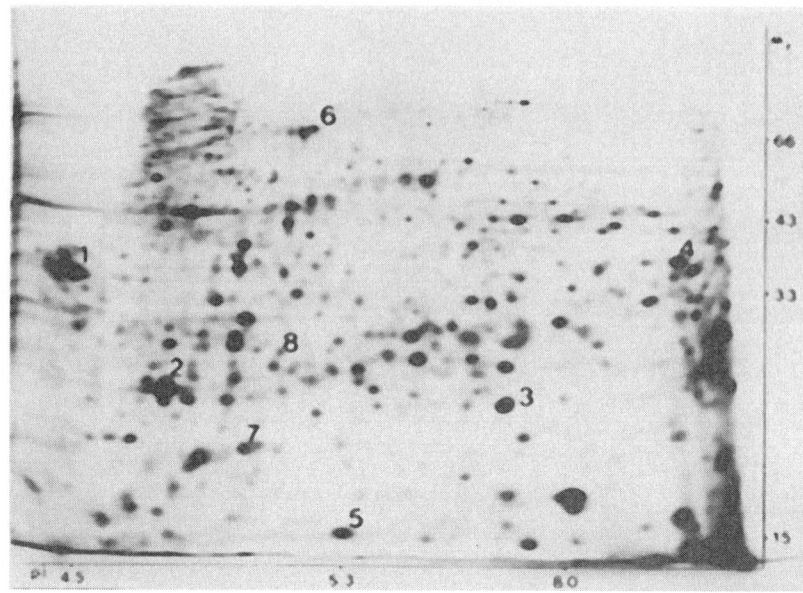

Fig. 4. Silver-stained 2-DE pattern of human myocardial proteins. A total of 50 µg protein were applied to the IEF gels. Calibration of pI and Mr was performed using internal marker proteins. Of the 100 proteins identified by amino acid analysis, sequencing and mass spectrometry, eight are indicated with numbers: *1*, tropomyosin; *2*, myosin light chain 1; *3*, α-crystallin b; *4*, glycerol aldehydephosphate dehydrogenase; *5*, fatty acid binding protein; *6*, serum albumin; *7*, malate dehydrogenase; *8*, heat-shock protein 27

the method has to be indicated. Human cardiac tropomyosin and human α-hemoglobin with pI values of 4.5 and 8.7, respectively, are well resolved under the conditions used here. By applying a low porosity cap gel at the cathode side of the separating gel, very basic proteins are stopped at the cap gel. Intensive staining without spot resolution at the cathode end of the gel indicates that a lot of basic proteins are within the investigated sample.

Calibration of molecular mass is done by marker proteins. An accuracy of 10% can be obtained; however, one should be aware that because of posttranslational modifications the molecular mass of a protein sometimes cannot be calculated from the

Fig. 5. Reproducibility of a sector of 2-DE gels from mouse brain. A total of 140 µg of protein were applied to the IEF gels. The proteins were stained with Coomassie brilliant blue R-250. A portion of the pattern of Fig. 3, containing creatine kinase and serum albumin, is shown. The same sample was applied 12 times

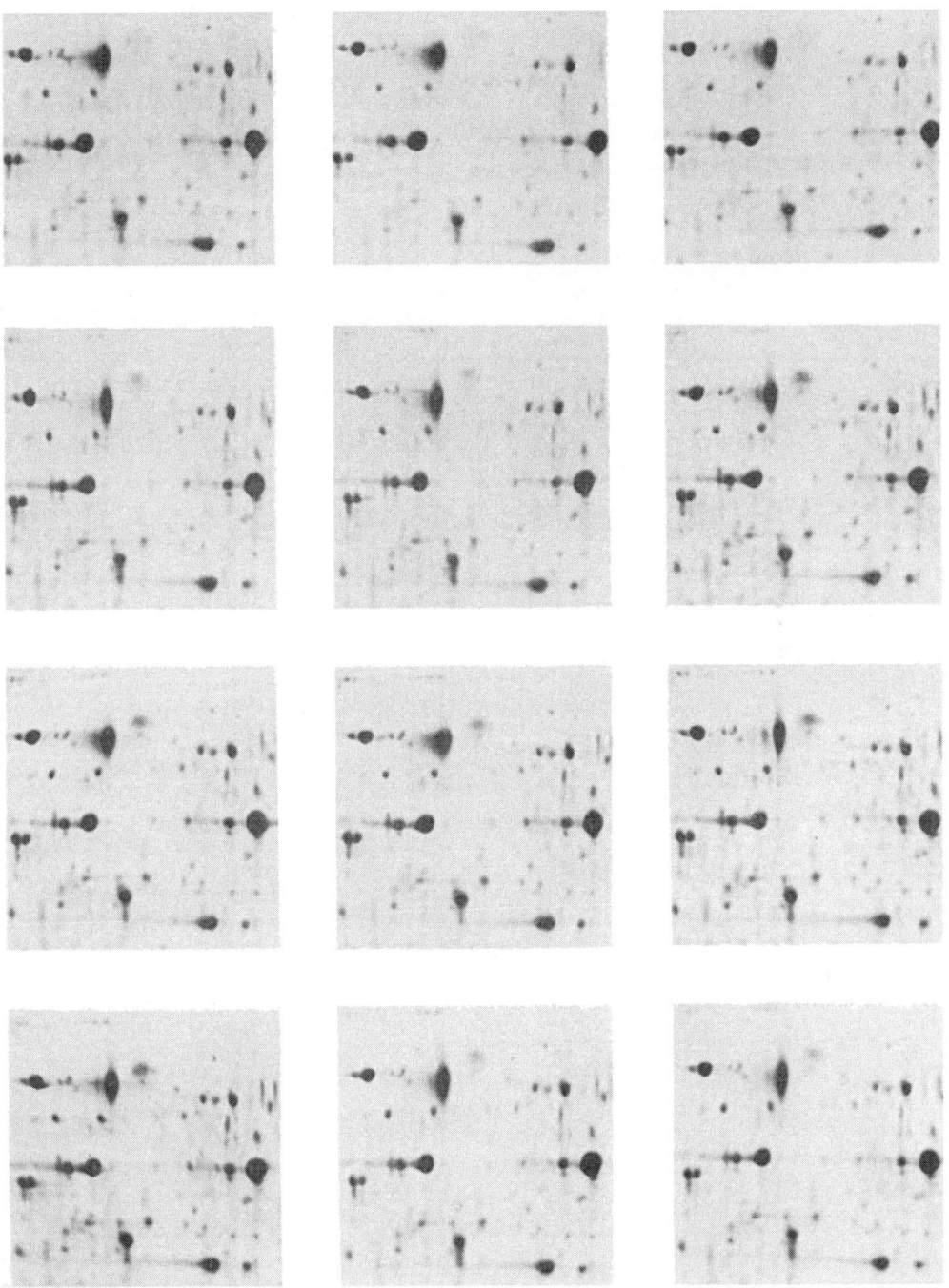

amino acid sequence. Marker proteins optimized for the 2-DE technique described here are available by BioRad (Munich, Germany). Optimal resolution is obtained for proteins in the molecular mass range of 14–66 kDa, shown in Fig. 4 by fatty acid binding protein and serum albumin. Good resolution is obtained for 10–100 kDa proteins.

13.11
Troubleshooting

IEF • Air bubbles within the IEF gels after polymerization: better degassing of the gel solution and longer polymerization times are required. One cause is a too high temperature during polymerization. Lower temperatures may be obtained by precooling the tubes at 4 °C for half an hour directly before use. Avoid exposure to sunlight during polymerization.

• Urea is crystallizing in the cathode solution during the IEF run: increase the room temperature. At temperatures below 20 °C urea crystallizes.

• Breaking of IEF at the cathode side indicates a cathodic drift effect. The following suggestions should alleviate the problem:

   – Clean the tubes with the procedure described above.

   – Use new gel solutions.

   – Decrease APS.

   – Decrease Vh product by decreasing each IEF step from 75 min to 60 min.

   – Sometimes the causes of cathodic drift effects are connected with the sample. Repeat sample preparation.

• The current is too high during IEF run: sealings are old, the screws are not tightened, or anode solution is dripping into the cathode solution. Tighten the screws and renew the cathode solution.

• Spots are not well separated or smearing occurs in the horizontal direction of the 2-DE gels: use new ampholytes and new gel solutions. The sample should be a clear solution. Some spots tend to streaking, e.g., actin always streaks in patterns of human myocardium.

• IEF gels are destroyed after thawing: freeze at –70 °C and thaw at room temperature.

- Membrane proteins are not migrating into the IEF gel: use 2% NP-40 or Chaps within the sample as well as within Sephadex and gel solutions.

- The current is too high during SDS-PAGE run: connections other than those between cathode and anode are present. Stop the run, remove the lid of the chamber and dry the protruding plastic part that divides anode and cathode.   **SDS-PAGE**

- The current is too low during SDS-PAGE run: the cathode and anode are not connected. Stop the run, remove the lid and fill electrode buffer into the anode part of the chamber so that the gel is covered by electrode buffer. Fill the cathode part to the same level as the anode part.

- Proteins below 20 kDa are not separated: porosity of the SDS-PAGE gels is too high or the pH value within the separation gel is too low. Variations of 0.1 pH unit have dramatic effects on the 2-DE patterns. Increase acrylamide concentration and check pH value of separation gel. Temperature also clearly effects the pH value of Tris buffers (decreasing pH with increasing temperature).

- The intensity of the spots is too low: **Silver staining**

  - Insufficient amount of protein in the sample. Increase the concentration of the sample or apply larger volumes to the IEF gels.

  - Silver staining is not working correctly due to old glutaraldehyde, washing steps that are too long, or incorrect pH of development solution. Use fresh glutaraldehyde for electron microscopy (Merck, Darmstadt, Germany), wash exactly as described and adjust the pH value of the developer exactly.

- Background is too high: pH value of the developer is too high, thiosulfate in the developer has been omitted. Another cause may be that the staining dishes have not been cleaned well enough. Clean them thoroughly using distilled water, 70% ethanol and Silvosol (Roth, FRG). Increase volume of solutions for silver staining.

- Even after a long development time only low intensity yellow spots appear: formaldehyde has been omitted within the developer.

- There are point streaks on the gels. Point streaks are thin but intensive vertical streaks up to 10 cm in length. Sometimes they occur in bundles of 10–20 streaks. Clean the glass plates of the SDS-PAGE gels thoroughly. Other possibilities are the use of alkylation procedures or the reduction by dithiothreitol within the incubation solution before the run of the second dimension.

2-DE Gels • Poor reproducibility: Train yourself to work exactly according to the instructions and in as little time as possible. The application of Sephadex, sample, overlay solution and anode solution should be finished within 15 min for eight samples. Incubation time of IEF gels should be kept constant. IEF gels should be applied to the second dimension directly after thawing. Use ready-made gel solutions.

• Dark clouds occur in the lower part of the gel: old ampholytes. Remove by fixation at 70 °C for 1 h. Use new gel solutions with new ampholytes.

• No high molecular mass proteins on the pattern: proteins are cleaved into peptides, e.g., by protease digestion. If sample is too old, use other samples. Change sample preparation for better prevention of degradation (other protease inhibitors). Biological material should· be kept frozen at –70 °C all the time especially before sample preparation.

• There are flags on top of the spots: increase polymerization time for the IEF gels up to 3 days.

References

Aebersold RH, Teplow DB, Hood LE, Kent SB (1986) Electroblotting onto activated glass: High efficiency preparation of proteins from analytical sodium dodecyl sulfate polyacrylamide gels for direct sequence analysis. J Biol Chem 261: 4229–4238

Aebersold R, Laevitt J (1990) Sequence analysis of proteins separated by polyacrylamide gel electrophoresis: Towards an integrated protein database. Electrophoresis 11: 517–527

Baker CS, Corbett JM, May AJ, Yacoub MH, Dunn MJ (1992) A human myocardial two-dimensional electrophoresis database: Protein characterisation by microsequencing and immunoblotting. Electrophoresis 13: 723–726

Deising H, Jungblut PR, Mendgen K, (1991) Differentiation-related proteins of the broad bean rust fungus Uromyces viciae-fabae, as revealed by high resolution two-dimensional polyacrylamide gel electrophoresis. ArchMicrobiol 155: 191–198

Eckerskorn C, Jungblut P, Mewes W, Klose J, Lottspeich F (1988) Identification of mouse brain proteins after two-dimensional electrophoresis and electroblotting by microsequence analysis and amino acid composition analysis. Electrophoresis 9: 830–838

Görg A, Postel W, Günther S (1988) The current state of two-dimensional electrophoresis with immobilized pH gradients. Electrophoresis 9: 531–546

Harrington MG, Lee KH, Yun M, Zewert T, Baley JE, Hood L (1993) Mechanical precision in two-dimensional electrophoresis can improve protein spot positional reproducibility. Appl Theor Electrophoresis 3: 347–353

Heukeshoven J, Dernick R (1985) Simplified method for silver staining of proteins in polyacrylamide gels and the mechanism of silver staining. Electrophoresis 6: 103–112

Jungblut P, Schneider W, Klose J. (1984) Quantitative analysis of two-dimensional electrophoretic protein patterns: Comparison of visual evaluation with computer-assisted evaluation. In: Neuhoff V (ed), Electrophoresis'84, Verlag Chemie GmbH, Weinheim, pp 301–303

Jungblut P, Klose J, (1985) Genetic Variability of proteins from mitochondria and mitochondrial fractions of mouse organs. Biochem Genet 23: 227–245

Jungblut P, Klose J (1986) Composition and genetic variability of Heparin-Sepharose CL-6B protein fractions obtained from the solubilized proteins of mouse organs. Biochem Genet 24: 925–939

Jungblut P, Prehm J, Klose J (1987) An attempt to resolve all the various proteins of a single human cell type by two-dimensional electrophoresis. J Biol Chem Hoppe-Seyler 367: 439

Jungblut P, Zimny-Arndt U, Klose J (1989) Composition and genetic variability of proteins from nuclear fractions of mouse (DBA/2J and C57BL/6J) liver and brain. Electrophoresis 10: 464–472

Jungblut PR, Seifert R (1990) Analysis by high-resolution two-dimensional electrophoresis of differentiation-dependent alterations in cytosolic protein pattern of HL-60 leukemic cells. J Biochem Biophys Meth 21: 47–58

Jungblut P, Dzionara M, Klose J, Wittmann-Liebold B, (1992) Identification of tissue proteins by amino acid analysis after purification by two-dimensional electrophoresis. J Prot Chem 11: 603–612

Jungblut P, Otto A, Zeindl-Eberhart E, Pleißner K-P, Knecht M, Regitz-Zagrosek V, Fleck E, Wittmann-Liebold B (1994) Protein composition of the human heart: The construction of a myocardial two-dimensional electrophoresis database. Electrophoresis 15: 685–707

Kaltschmidt E, Wittmann HG (1970) Ribosomal proteins. VII Two dimensional polyacrylamide gel electrophoresis for fingerprinting of ribosomal proteins. Anal Biochem 36: 401–412

Kamp RM, Wittmann-Liebold B (1984) Purification of Escherichia coli 50 S ribosomal proteins by high performance liquid chromatography. FEBS Lett 167: 59–63

Klose J (1975) Protein mapping by combined isoelectric focusing and electrophoresis in mouse tissues. A novel approach to testing for induced point mutations in mammals. Humangenetik 26: 211–234

Klose J, Kobalz U (1995) Two-dimensional electrophoresis of proteins: An updated protocol and implications for a functional analysis of the genome. Electrophoresis 16: 1034–1059

Knecht M, Regitz-Zagrosek V, Pleissner K-P, Jungblut P, Hildebrandt A, Fleck E (1994) Characterization of the myocardial protein composition in dilated cardiomyopathy by two-dimensional gel electrophoresis. Europ Heart J 15 Supplement D: 37–44

Merril CR, Switzer RC, Van Keuren ML (1979) Trace polypeptides in cellular extracts and human body fluids detected by two-dimensional electrophoresis and a highly sensitive silver stain. Proc Natl Acad Sci USA 76: 4335–4339

Nokihara K, Morita N, Kuriki T (1992) Applications of an automated apparatus for two-dimensional electrophoresis, Model TEP-01, for microsequence analyzes of proteins. Electrophoresis 13: 701–707

O'Farrell PH (1975) High resolution two-dimensional electrophoresis of proteins. J Biol Chem 250: 4007–4021

Otto A, Benndorf R, Wittmann-Liebold B, Jungblut P, (1994) Identification of proteins on two-dimensional gels for the construction of a human heart 2-DE database. J Prot Chem. 13: 478–480

Pappin DJC, Hojrup P, Bleasby AJ (1993) Rapid identification of proteins by peptide-mass fingerprinting. Curr Biol 3: 327–332

Prehm J, Jungblut P, Klose J (1987) Analysis of two-dimensional electrophoretic protein patterns using a video camera and a computer II. Adaption of automatic spot detection to visual evaluation. Electrophoresis 8: 562–572

Rabilloud T, Vuillard L, Gilly C, Lawrence JJ (1994) Silver staining of proteins in polyacrylamide gels: A general overview. Cell Mol Biol 40: 57–75

Raymond S (1964) Acrylamide gel electrophoresis. Ann NY Acad Sci 121: 350–365

Scheele GA (1975) Two-dimensional gel analysis of soluble proteins. Characterization of guinea pig exocrine pancreatic proteins. J Biol Chem. 250: 5375–5385

Smithies O, Poulik MD (1956) Two-dimensional electrophoresis of serum proteins. Nature 177: 1033

Stegemann H, Francksen H, Macko V (1973) Potato proteins: Genetic and physiological changes, evaluated by one- and two-dimensional PAA-gel-techniques. Z Naturforsch 28c: 722–732

Vandekerckhove J, Bauw G, Puype M, Van Damme J, Van Montagu M (1985) Protein-blotting on polybrene-coated glass-fiber sheets. A basis of acid hydrolysis and gas-phase sequencing of picomole quantities of protein previously separated on SDS-polyacrylamide gel. Eur J Biochem. 152: 9–19

Zeindl-Eberhart E, Jungblut P, Otto A, Rabes HM (1994) Characterization of tumor-associated protein variants during hepatocarcinogenesis in the rat. J Biol Chem 269: 14589–14594

Semi-Dry Blotting onto Hydrophobic Membranes

P. Jungblut

14.1
Introduction

Polyacrylamide is a very good matrix for the separation of proteins, but it is a poor matrix for their characterization, with the exceptions of molecular mass and isoelectric point estimations. For most proteins elution from gels is accompanied with losses of up to 90%. The introduction of nitrocellulose as a matrix onto which proteins can be blotted and on which the proteins are immobilized has allowed a wide range of characterization reactions. The first membrane-immobilized proteins were characterized by immunostaining (Towbin et al. 1979; Burnette 1981). Other overlay techniques are lectin staining for the identification of glycoproteins (Clegg 1982), radioactive calcium overlays (Maruyama et al. 1984), or GTP overlays for the identification of GTP-binding proteins (McGrath et al. 1984). However, protein-chemical investigations cannot be performed within an acrylamide gel and nitrocellulose membrane is not compatible with automated NH_2-terminal sequencing. Therefore, an important advance in the combination of two-dimensional electrophoresis (2-DE) and protein-chemical methods was the development of inert membranes onto which proteins can be blotted with high efficiency. The first sequencer-stable membranes were glass fiber membranes modified either with positively charged (Vandekerckhove et al. 1985; Aebersold et al. 1986) or hydrophobic, uncharged (Eckerskorn et al. 1988a) groups, or pure organic polymers such as polyvinylidene difluoride (PVDF) (Matsudaira 1987) or polypropylene (PP) (Eckerskorn and Lottspeich 1990; Jungblut et al. 1990). For different hydrophobic membranes and proteins the blotting efficiency has been tested, the blotting parameters optimized and a blotting mechanism proposed (Jungblut et al. 1990). The suitability of these membranes for sequencing was compared by Eckerskorn and Lottspeich (1993).

Table 1. Parameters influencing the transfer of proteins out of SDS gels

	Parameter	Transfer improved by
Proteins	Mr	Decrease
SDS-gels	SDS concentration	Increase
	Acrylamide composition	Decrease
	Ionic strength after electrophoresis	Decrease
Blotting buffer	Ionic strength	Decrease
	Methanol concentration	Decrease
	pH	Increase
Blotting conditions	Current/area	Increase
	Time[a]	Increase

[a]A better transfer by increasing the time is only obtained as long as proteins are loaded with SDS. By addition of SDS to the blotting buffer a further increase in time may be useful.

The first blotting experiments were performed in blotting tanks. Semi-dry blotting was first reported by Kyhse-Andersen in 1984. It has the advantage of easier performance, lower buffer consumption and therefore less contamination.

Blotting efficiency depends on the efficiency of the transfer from the gel to the membrane and on the efficiency of the binding of the protein to the membrane. In Table 1, the parameters influencing transfer of proteins out of gels and suggestions for improving the transfer are listed. Table 2 lists the parameters that influence membrane binding and suggestions for improving binding.

Table 2. Parameters influencing binding of proteins on hydrophobic mebranes

	Parameter	Binding improved by
Proteins	Mr	Increase
SDS gels	SDS concentration	Decrease
	Ionic strength after electrophoresis	Increase
Blotting buffer:	Ionic strength	Increase
	Methanol concentration	Increase
	pH	Decrease
Blotting conditions	Current/area	Decrease
	Time	No effect[a]
Membrane type:	Hydrophobicity	Increase

[a]Increasing the blotting time does not remove bound protein from the mebrane.

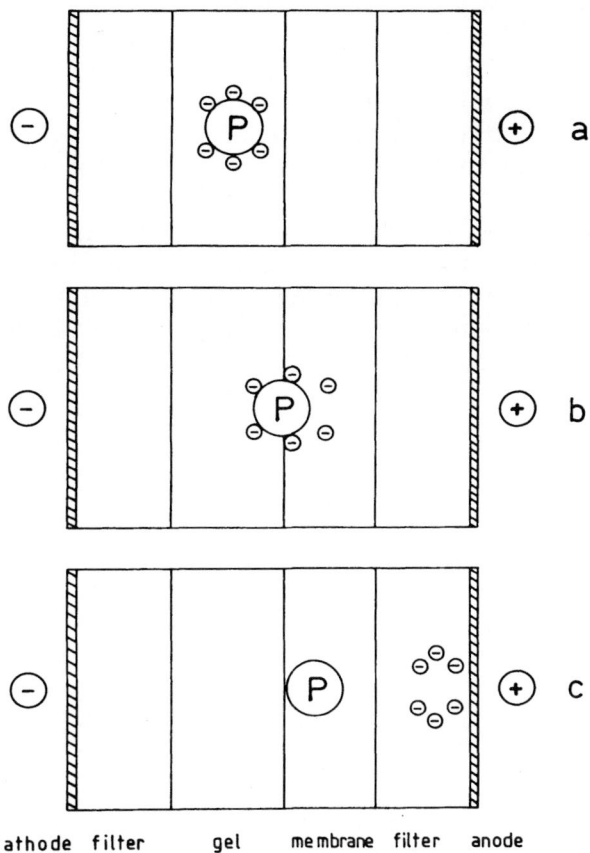

cathode filter   gel   membrane filter   anode

Fig. 1a–c. Blotting mechanism. **a** Within the SDS gel the protein (*P*) is loaded with SDS (*dash in small circles*). **b** The protein-SDS complexes migrate out of the gel onto the membrane. SDS is partly stripped off from the protein during this migration. **c** The protein binds with the SDS-free parts of its amino acids to the membrane by hydrophobic interactions; subsequently, SDS is completely stripped off from the protein. The protein binds to the membrane and SDS moves to the anode. For further details, see text

The blotting mechanism described in Fig. 1 is consistent with the parameter influences described in the tables. At the beginning of a blotting experiment the protein is situated within the gel and is loaded with SDS. The negative charge of SDS pulls the SDS-protein complex towards the anode. During the resulting migration SDS is stripped off from the protein. At the moment the protein reaches the hydrophobic membrane some of the hydrophobic binding sites of the protein must be free of SDS to allow hydrophobic interaction with the membrane. The protein binds to the membrane; the remaining SDS is also stripped off from the protein and migrates to the anode, whereas the protein remains bound on the membrane. This mechanism is consistent with the observation that an increase only of the blotting time will not result in loss of protein from the membrane. High molecular mass proteins can lose their SDS before they have left the gel. An increase of blotting time will not increase blotting efficiency, because proteins without negative charge cannot move out of the gel. An increase of blotting time makes sense only by adding SDS to the cathode blotting buffer. Low molecular mass proteins, which migrate too fast, have not had some of the SDS molecules stripped off at the moment they reach the membrane. They cannot build up their hydrophobic interactions and thus pass the membrane without binding.

14.2
Flow Chart

Blotting should be performed directly after SDS-PAGE. Equilibration of the gels in blotting buffer is not necessary if 15% acrylamide gels, 1.5 mm thick and prepared according to the method of Laemmli (1970), are used.

- SDS-PAGE run

- Preparation of the blotting sandwich up to the three layers of anode filter papers

- Stop of SDS-PAGE run

- Completing the blotting sandwich with membrane, gel and cathode filter papers

- Blotting run

- Staining

14.3
Materials

Equipment
- semi-dry transfer cell: e.g., Sartoblot II (Sartorius, Heidelberg, Germany, graphite); Trans-Blot SD (BioRad, Munich, Germany, cathode: stainless steel, anode: Pt coated Ti); Pharmacia-LKB (graphite; Pharmacia, Freiburg, Germany); Schleicher and Schuell (Carboglass; Schleicher and Schuell, Dassel, Germany)

- filter paper, Schleicher and Schuell, 426 892 003

- PVDF: Immobilon (Millipore, Eschborn, Germany); Transblot (BioRad, Munich Germany); Problot (Perkin Elmer Applied Biosystems, Weiterstadt, Germany) silanized glass fiber (SGF): Glassybond (Biometra, Göttingen, Germany)

- gels: 1.5 mm 2-DE-gels

- deionized water

- scissors

- gloves

- glass rod, pipette

- dishes

- PVDF   Solutions

- methanol (Merck, Darmstadt, Germany; for PVDF)

- acetone (for Glassybond)

- blotting buffer: 50 mM borate/20% methanol (pH 9.0). For 1 l: add 3.09 g boric acid (Merck, Darmstadt, Germany), 750 ml deionized water, 200 ml methanol, and 18.6 ml 1 N NaOH. Bring final volume to 1 l with deionized water.

14.4
Procedure

1. Prepare membranes 30 min before the end of 2nd dimension.

1a. PVDF: dip in methanol for 20–30 s

1b. SGF: dip in acetone for 20–30 s

Shake membrane in blotting buffer: 3 × 5 min

2. Prepare blotting chamber by washing it with deionized water. Graphite electrodes should stay damp until the beginning of the blotting process.

3. Prepare the sandwich on the anode side **5 min before the end of 2nd dimension:** Roll three layers of filter paper individually onto the anode. The filter paper should extend about 3 mm on every side of the blotting gel. Expel air bubbles by forcefully rolling a pipette across the filter paper.

4. The membrane is laid carefully upwards. Do not roll the membrane. The membrane must not be allowed to dry before the gel is placed on it.

5. When **the 2nd dimension is finished,** the 2-DE-gel is laid directly on the membrane. Three layers of filter paper are laid individually on the gel. Each layer is pressed by rolling in order to remove air bubbles. The sandwich should not be too damp. There should be no liquid between the gel and the membrane.

6. Close the blotting chamber. If a chamber with graphite electrodes is used, place 1 kg of weight on the chamber to weigh it down.

7. Initial blotting conditions: 1 mA/cm^2 (i.e., 55 mA for 55 cm^2 gels and 110 mA for 110 cm^2 gels) for 3 h; voltage is limited to 100 V. During blotting, voltage increases from 6 V to 20 V.

14.5
Unspecific Staining of Proteins on Hydrophobic Membranes

14.5.1
Coomassie Brilliant Blue R-250

Coomassie brilliant blue R-250 staining of hydrophobic membranes is compatible with sequencing. Sensitivity is higher than with Poinceau S and comparable with that of sulforhodamine B. The stain is not removable without loss of protein. The following procedure is based on that of Eckerskorn et al. (1988). It is not recommended for nitrocellulose membranes.

Solutions   – staining solution: 0.1% Serva blue R-250/40% methanol/10% acetic acid. For 500 ml: 1 g Serva blue R- 250, 400 ml methanol, 100 ml acetic acid (Merck, Darmstadt, Germany). Bring to final volume of 500 ml with deionized water.

  – PVDF destaining solution: 40% methanol/10% acetic acid. For 1 l: 400 ml methanol, 100 ml acetic acid (Merck, Darmstadt, Germany), 500 ml deionized water.
SGF destaining solution: 30% methanol/10% acetic acid. For 1 l: 300 ml methanol, 100 ml acetic acid (Merck, Darmstadt, Germany), 600 ml deionized water.

Procedure   For 6.5 × 7.5 cm blots, about 100 ml of staining and destaining solutions per step are sufficient.

1. Carefully shake membrane in staining solution for 5 min.

2. Carefully shake membrane in destaining solution for 3 × 5 min

3. Dry the membrane by hanging it up or by laying it onto several layers of filter papers.

14.5.2
Sulforhodamine B

Sulforhodamine B staining is compatible with sequencing and with matrix-assisted laser/desorption ionization mass spectrometry. The sensitivity is comparable to that of Coomassie brilliant blue R-250. The following procedure is based on that of Coull and Pappin (1990). It cannot be used for nitrocellulose membranes.

– staining solution: 30% methanol, 0.2% (v/v) acetic acid, **Solution** 0.005% (w/v) sulforhodamine B. For 500 ml: 150 ml methanol (Merck, Darmstadt, Germany, uvasol), 1 ml acetic acid, 25 mg sulforhodamine B. Fill up to 500 ml with deionized water.

For a 6.5 × 7.5 cm membrane, about 100 ml of staining solution is **Procedure** sufficient.

1. Wash membranes in 100 ml deionized water directly after blotting to remove buffer.

2. Completely dry membranes (room temperature overnight, or in a desiccator).

3. Shake dried membrane in staining solution for 1–2 min.

4. Destain membrane in deionized water for several seconds.

14.5.3
Poinceau S

Poinceau S staining is reversible and can be used on nitrocellulose and on hydrophobic membranes. It is used for localization of proteins before immunostaining. Due to its low sensitivity, it is not advantageous for sequencing and mass spectrometry. The following procedure is based on that of Salinowich and Montelaro (1986).

– staining solution: 0.5% Poinceau S, 1% acetic acid. Dissolve **Solution** 0.5 g Poinceau S in 100 ml of 1% acetic acid.

For 6.5 × 7.5 cm membranes, 100 ml of staining solution are suf- **Procedure** ficient.

1. After blotting, stain the membrane for 5 min in staining solution.

2. Destain the background in distilled water for 2–3 min.

3. Mark the protein spots.

4. Completely destain, including the protein spots, by washing in distilled water for 5–10 min.

14.5.4
Immunostaining

Many specific stains for protein bound onto membranes have been described. The following immunostaining procedure is easy to perform and the resulting red stain can be observed visually and is stable over a long period of time. The blocking and staining steps can be used for nitrocellulose and PVDF membranes. This method is based on that of Pluzek and Ramlau (1988).

Materials
- glass or plastic trough the same size as the membrane to be stained
- shaker
- Tris/NaCl: 100 mM Tris-HCl (pH 7.5), 135 mM NaCl
- phosphate-buffered saline (PBS, pH 7.4) : 20 ml 1 M KH_2PO_4, 80 ml 1 M K_2HPO_4, 45 g NaCl. Bring volume to 5 l with deionized water.
- PBS/Tween: 1 l PBS (pH 7.4), 3 g Tween-20
- PBS/Tween/DM 3%: 100 ml PBS/Tween, 3 g non-fat dry milk (DM)
- solution 1: 0.8 M Tris (9.6912 g Tris; final volume 100 ml with double distilled water)
- solution 2: 0.4 N HCl (40 ml 1 N HCl, 60 ml double distilled water)
- solution 3: 0.048 M $MgCl_2$ (2.033 g $MgCl_2 \times 6H_2O$; final volume 100 ml with double distilled water)
- substrate buffer: 50.0 ml solution 1, 53.6 ml solution 2, 4.1 ml solution 3. Add 100 ml of double distilled water. Final volume:207.7 ml; pH 8.0
- substrate

Note: The substrate must be mixed directly before use. The reaction must be performed in the dark.

- naphthol (Sigma, St. Louis, USA) (store at –20 °C): 0.4 mg/ml substrate buffer
- Fast Red (Sigma, St. Louis, USA) (store at –20 °C): 6.0 mg/ml substrate buffer

Mix identical volumes of each solution.

- primary antibody. The animal source and the Ig class of the antibody should be known. Dilute the antibody 1:20 up to 1:10|000, depending on the antibody, in PBS/Tween/DM 1% (for example, rabbit serum against human myoglobin: dilution 1:10|000)
- secondary antibody conjugated with alkaline phosphatase. The antibody is from a species different from the one used for the primary antibody and must bind to the primary antibody. Dilute the antibody with PBS/Tween/DM 1% (for example, alkaline phosphatase goat anti-rabbit IgG Fc, dilution 1:5000).

All steps are performed on the shaker.

Saturation of the blotting membrane

1. After blotting, wash the membranes 3×10 min in Tris/NaCl to remove salts, SDS and glycine.

2. Block the membranes by placing them in PBS/Tween/DM. Incubate for at least 1 h. Overnight blocking is also possible.

3. Wash the membranes for 3×10 min in PBS/Tween.

4. Dry the membranes or use them directly for immunostaining.

Immunostaining procedure

1. Incubate the saturated and washed membranes in a solution of the primary antibody overnight at 4°C.

   Note: Dried hydrohobic membranes such as PVDF membranes have to be wetted by dipping them in methanol directly before incubation with antibody.

2. Wash the membranes 3× for 10 min in PBS/Tween.

3. Incubate the secondary antibody with the membrane for 1 h at room temperature.

4. Wash the membranes 3×10 min in PBS/Tween. Continue washing with water until all foam is removed.

5. Preincubate substrate 10 min in substrate buffer.

6. substrate incubation:

Add the substrate (naphthol and Fast Red) in substrate buffer to the membranes and visually control staining of the proteins. When the desired staining intensity is reached, stop the reaction with deionized water. The membranes are dried in filter paper.

14.6
Results

Blotting onto hydrophobic membranes has been used for the identification of 2-DE separated proteins from different sources. Figure 2 shows part of a mouse brain protein pattern on a PVDF membrane. In this case for protein separation a high performance 2-DE system ($23 \times 30 \times 0.15$ cm) was used. The membrane was stained with Coomassie brilliant blue R-250. Some of the proteins were identified by direct NH_2-terminal sequencing

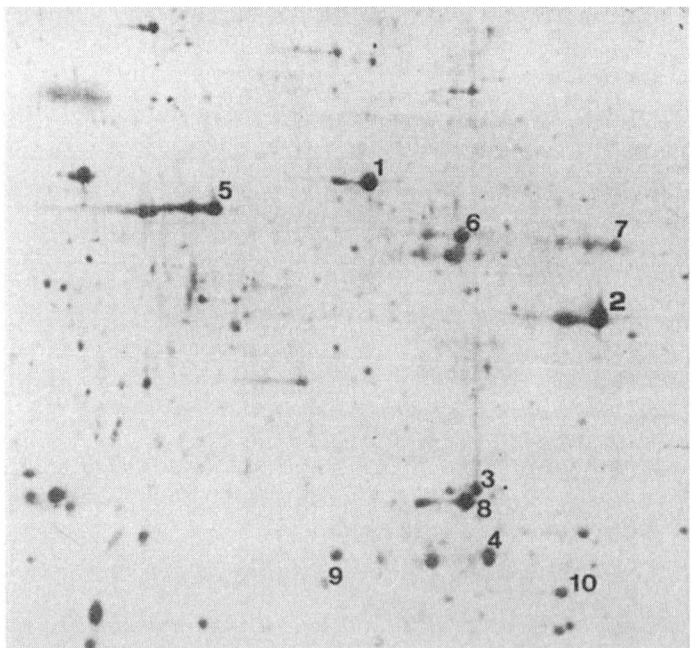

Fig. 2. Portion of a high performance 2-DE pattern of mouse brain proteins blotted onto PVDF membrane. The proteins were stained with Coomassie brilliant blue R-250. Of the 43 proteins identified by amino acid analysis (Jungblut et al. 1992), ten are indicated: *1*, creatine kinase; *2*, glycerol aldehyde phosphate dehydrogenase; *3*, carbonic anhydrase; *4*, triose-phosphate isomerase; *5*, actin; *6*, fructose bisphosphatase; *7*, pyruvate dehydrogenase; *8*, phospho glycerate mutase B; *9*, thymidine kinase; *10*, glutathione transferase

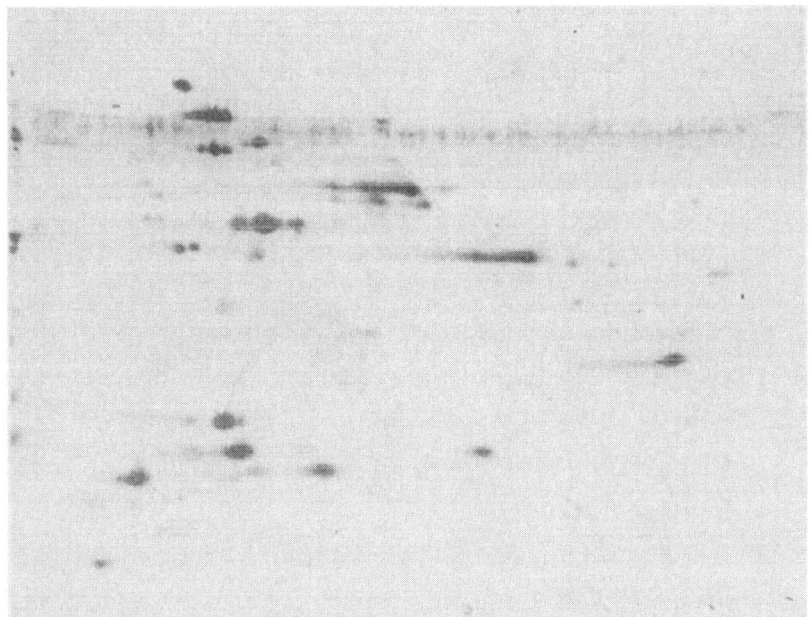

Fig. 3. Immunoblot of a human serum against *Borrelia burgdorferi* proteins. *Borrelia burgdorferi* proteins were separated with a small gel 2-DE technique. The 2-DE pattern was blotted onto a PVDF membrane. The serum of a patient suffering from borreliosis was used as the primary antibody in the immunostaining procedure. Naphthol and Fast Red were used as substrates

(Eckerskorn et al. 1988) and others by amino acid analysis (Jungblut et al. 1992). Some examples of identified proteins are marked on the figure. The same blotting technique was also used for internal sequencing of human myocardial proteins (Jungblut et al. 1994) and for the identification of a rat tumor-associated protein (Zeindl-Eberhart et al. 1994). For matrix-assisted laser desorption/ionization mass spectrometry, sulforhodamine staining was preferred and 15 proteins were identified on a myocardial 2-DE protein map (Thiede et al. 1996). In this case posttranslational modifications were also detected.

Antigens of microorganisms can be detected by immunoblotting. *Borrelia burgdorferi* proteins have been separated using a small gel 2-DE technique and antibodies against *Borrelia burgdorferi* proteins have been looked for in the serum of a patient. The serum was used as the primary antibody and the substrates as described above were used. The immunoblot is shown in Fig. 3. The patient had antibodies against about 40 proteins of *Borrelia burgdorferi*.

14.7
Troubleshooting

- Drop-like smear (swirls) on the membranes: The blotting sandwich was too wet. Blotting buffer has to be pressed out better during assembly of the sandwich.

- Voltage increases drastically and current is too low: The blotting sandwich is too dry. Check to make sure that the hydrophobic membrane was prewetted with methanol.

- High molecular mass proteins remaining in the gel.

  - Decrease methanol concentration within the blotting buffer

  - Decrease ionic strength in the blotting buffer

  - Increase current/area

  - Use SDS within the cathode blotting buffer and increase the blotting time

- Low molecular mass proteins do not bind to the membrane

  - Increase methanol concentration within the blotting buffer

  - Increase ionic strength within the blotting buffer

  - Decrease current/area

  - Use a more hydrophobic membrane

References

Aebersold RH, Teplow DB, Hood LE, Kent SB (1986) Electroblotting onto activated glass: High efficiency preparation of proteins from analytical sodium dodecyl sulfate polyacrylamide gels for direct sequence analysis. J Biol Chem 261: 4229–4238

Burnette WN (1981) Western blotting: Electrophoretic transfer of proteins from sodium dodecyl sulfate-polyacrylamide gels to unmodified nitrocellulose and radiographic detection with antibody and radioiodinated protein. Anal Biochem 112: 195–203

Clegg JCS (1982) Glycoprotein detection in nitrocellulose transfers of electrophoretically separated protein mixtures using concanavalin A and peroxidase: Application to arenavirus and flavivirus proteins. Anal Biochem 127: 389–394

Coull JM, Pappin DJC (1990) A rapid fluorescent staining procedure for proteins electroblotted onto PVDF membranes. J Prot Chem 9: 259–260

Eckerskorn C, Jungblut P, Mewes W, Klose J, Lottspeich F, (1988) Identification of mouse brain proteins after two-dimensional electrophoresis and electroblotting by microsequence analysis and amino acid composition analysis. Electrophoresis 9: 830–838

Eckerskorn C, Mewes W, Goretzki H, Lottspeich F (1988) A new siliconized glass-fiber as support for protein-chemical analysis of electroblotted proteins. Eur J Biochem 176: 509–519

Eckerskorn C, Lottspeich F (1990) Combination of two-dimensional gel electrophoresis with microsequencing and amino acid composition analysis: Improvement of speed and sensitivity in protein characterization. Electrophoresis 11: 554–561

Eckerskorn C, Lottspeich F (1993) Structural characterization of blotting membranes and the influence of membrane parameters for electroblotting and subsequent amino acid sequence analysis of proteins. Electrophoresis 14: 831–838

Jungblut P, Eckerskorn C, Lottspeich F, Klose J (1990) Blotting efficiency investigated by using two-dimensional electrophoresis, hydrophobic membranes and proteins from different sources. Electrophoresis 11: 581–588

Jungblut P, Dzionara M, Klose J, Wittmann-Liebold B, (1992) Identification of tissue proteins by amino acid analysis after purification by two-dimensional electrophoresis. J Prot Chem 11: 603–612

Jungblut P, Otto A, Zeindl-Eberhart E, Pleißner K-P, Knecht M, Regitz-Zagrosek V, Fleck E, Wittmann-Liebold B (1994) Protein composition of the human heart: The construction of a myocardial two-dimensional electrophoresis database. Electrophoresis 15: 685–707

Laemmli UK (1970) Cleavage of structural proteins during the assembly of the head of bacteriophage T4. Nature 227: 680–685

Khyse-Andersen J (1984) Electropblotting of multiple gels: A simple apparatus without buffer tank for rapid transfer of proteins from polyacrylamide to nitrocellulose. J Biochem Biophys Meth 10: 203–209

Maruyama K, Mikawa T, Ebashi S (1984) Detection of calcium binding proteins by ^{45}Ca autoradiography on nitrocellulose membrane after sodium dodecyl sulfate gel electrophoresis. J Biochem 95: 511–519

Matsudaira P (1987) Sequence from picomole quantities of proteins electroblotted onto polyvinylidene difluoride membranes. J Biol Chem 262: 10035–10038

McGrath JP, Capon DJ, Goeddel DV, Levinson AD (1984) Comparative biochemical properties of normal and activated human ras p21 protein. Nature 310: 644–649

Pluzek K-J, Ramlau J (1988) In: Bjerrum OJ, Heegaard NHH(eds) Handbook of immunoblotting of proteins, vol 1. CRC Press, Inc, Boca Raton, FL: 177

Salinowich O, Montelaro R (1986) Reversible staining and peptide mapping of proteins transferred to nitrocellulose after separation by sodium dodecylsulfate-polyacrylamide gel electrophoresis. Anal Biochem 156: 341–347

Towbin H, Staehelin T, Gordon J (1979) Electrophoretic transfer of proteins from polyacrylamide gels to nitrocellulose sheets: Procedure and some applications. Proc Natl Acad Sci USA 76: 4350–4354

Vandekerckhove J, Bauw G, Puype M, Van Damme J, Van Montagu M (1985) Protein-blotting on polybrene-coated glass-fiber sheets. A basis of acid hydrolysis and gas-phase sequencing of picomole quantities of protein previously separated on SDS-polyacrylamide gel. Eur J Biochem 152: 9–19

Thiede B, Otto A, Zimny-Arndt U, Müller E-C, Jungblut P (1996) Identification of human myocardial proteins separated by two-dimensional electrophoresis with matrix-assisted laser desorption/ionization mass spectrometry. Electrophoresis 17:588–599

Zeindl-Eberhart E, Jungblut P, Otto A, Rabes HM (1994) Characterization of tumor-associated protein variants during hepatocarcinogenesis in the rat. J Biol Chem 269: 14589–14594

Part V
Analysis of Amino Acids

Chapter 15

Highly Sensitive Amino Acid Analysis

R. M. KAMP

15.1
Introduction

Amino acid analysis has found wide application in many areas of research and industry such as biochemistry, biotechnology, diagnostic, neurobiology and quality control. Amino acids occur as free molecules in physiological samples and can be determined directly in the investigated sample. Determination of proteins and peptides is possible after total acidic or alkaline hydrolysis of the sample. Proteins include 20 α-amino acids, and more than 150 amino acids and their derivatives have been identified in different physiological samples.

Amino acid analysis was first introduced by Stanford Moore and Wiliam Stein at Rockefeller University in New York. Amino acids were separated on a sulfonated cation-exchange column in 4–6 h and the detection limit was 10 nmol after postcolumn derivatization with ninhydrin (Spackman 1958).This method offers the highest reproducibility and accuracy. In recent years high performance liquid chromatography (HPLC) has found wide application in amino acid analysis using precolumn derivatization with fluorescent reagents and separation on reversed phase columns. The major advantages of this method are fast and highly sensitive measurements.

15.2
Application

15.2.1
Amino Acid Analysis Using Precolumn Derivatization

Over the past 20 years, the use of HPLC in biomedical research has greatly increased, and the rapid development of new reversed phase chromatography and ion-exchange columns have allowed the application of HPLC to the determination of amino acids.

HPLC AMINO ACID ANALYSIS

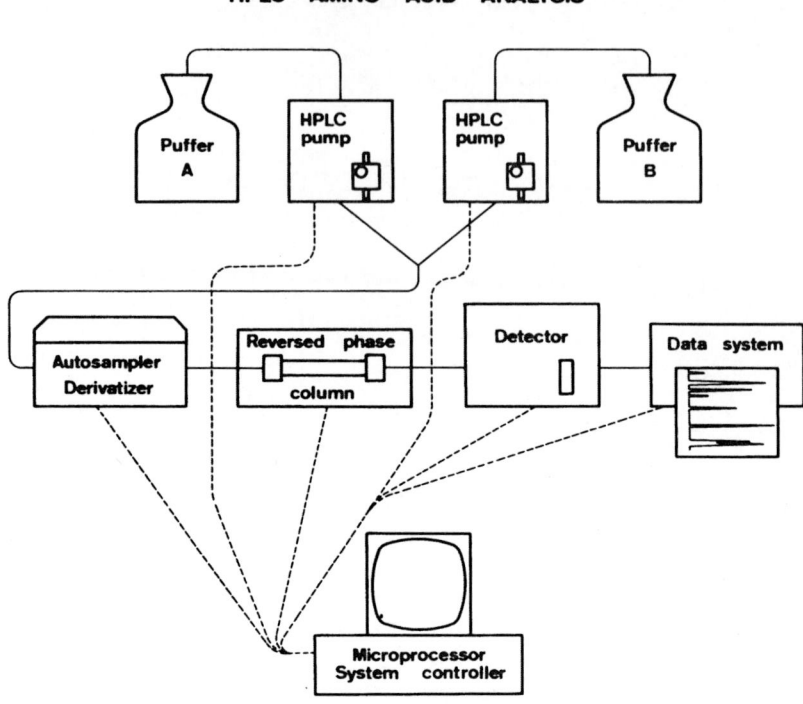

Fig. 1. HPLC amino acid analyzer with precolumn derivatization

Modern HPLC instrumentation and separation columns have resulted in increased sensitivity, from nanomole to picomole or femtomole levels, and decreased separation times, from hours to minutes, maintaining high resolution. The advent of HPLC and development of precolumn derivatization techniques offer an alternative to established methods. Figure 1 shows the scheme of the HPLC amino acid analyzer with precolumn derivatization. Different reagents can be used for amino acid analysis with precolumn derivatization. These include:

- ortho-phthaldialdehyde (OPA)

- 9-fluorenylmethyl chloroformate (FMOC-Cl)

- 4-chloro-7-nitro-2,1,3-benzoxadiazole (NBD-Cl)

- 4-(dimethylamino)azobenzene-4-sulfonyl   chloride   (dabsyl chloride)

- 6–aminoquinolyl-N-hydroxy-succinimidyl carbamate (AQC)

In 1971, Roth described the reaction of OPA with amino acids as an alternative to the classical ninhydrin method (Roth 1972). OPA derivatives are detectable by fluorescence at low picomole or femtomole levels. OPA reacts with primary amino acids in the pH range of 9–10 and forms fluorescent isoindoles. The OPA method provides high reproducibility and can be used for routine analysis. One disadvantage of OPA direct derivatization is that only primary amino acids can be determined. However, there are several ways to avoid this problem. Secondary amino acids can be detected using FMOC, dabsyl chloride, or phenylisothiocyanate. These reagents react with primary and secondary amino acids. OPA derivatives are stable for 1–2 min but degenerate quickly thereafter. For the best reproducibility of the OPA reaction, a fully automated system should be used.

OPA derivatization

FMOC was introduced as an amino acid derivatization reagent by Einarsson in 1983. It allows sensitive amino acid analysis using fluorescence detection. FMOC is highly fluorescent and hydrolyzes to the alcohol. Any excess FMOC by-products after reaction with amino acids must be removed before separation on an HPLC system. One method is extraction of by-products using pentane, hexane or mixtures of pentane and ethyl acetate. Extraction with pentane gives relatively good removal of FMOC alcohol, but it also removes undefined quantities of amino acids. The second method relies on the reaction with adamantane. FMOC reacts with adamantane to form a hydrophobic complex, which is eluted at the end of the chromatogram and hence does not hinder the identification of amino acid derivatives. FMOC derivatization is sensitive and easy to perform and allows the determination of primary and secondary amino acids. A disadvantage of the method is that it is not suitable for tryptophan or cysteine determination: the response of the FMOC derivative of tryptophan or cysteine is low due to intramolecular quenching of fluorescence. Although, these derivatives can be detected in the UV range, the sensitivity is very low.

FMOC derivatization

Precolumn derivatization with NBD-Cl can be applied as a complementary method to OPA for the determination of secondary amines (Carisano 1985). In medical research it is often necessary to selectively determine proline and hydroxyproline in serum, collagen hydrolyzate, or tissue and tissue extracts; for example, in the diagnosis of chronic kidney failure. Palmerini (1985) described the first selective determination of secondary amines. A modified method, described below, can be used. Amino acids react with the OPA/NBD-Cl mixture in the following manner. OPA reacts rapidly with primary amino acids and, without 2-

Selective determination of proline and hydroxyproline using NBD-Cl

mercaptoethanol, does not form fluorescent derivatives. NBD-Cl simultaneously reacts with the secondary amines, and their derivatives are detected using fluorescence. The selective determination of secondary amino acids is rapid and simple. Separation can be completed in 5 min.

The NBD-Cl derivatives are stable, but their sensitivity, compared with OPA or FMOC derivatives, is approximately 100 times lower. Neverless, it is possible to detect NBD-amino acids at low picomole levels.

Dabsyl chloride   Dabsyl chloride was introduced by Chang in 1981 for amino acid analysis by manual derivatization and HPLC separation (Chang 1983). Dabsyl chloride reacts with primary and secondary amines to form orange-coloured derivatives which are detectable at 436 nm and are stable for months. This method permits the simultaneous determination of primary and secondary amines.The sensitivity is at the low picomole level. The disadvantage is the relatively short column life. An excess of dabsyl chloride reagents are continually injected onto the column together with the derivatized sample which sticks to the columns thus causing their deterioration.

AQC derivatization   In 1994 Millipore (USA) introduced 6-aminoquinolyl-N-hydroxysuccinimidyl carbamate (AQC) for rapid and simple amino acid analysis. This reagent forms stable derivatives with primary and secondary amino acids. The derivatives are easily separated by reversed phase HPLC in less than 35 min. AQC is consumed during the reaction to form aminoquinoline (AMQ), which has different spectral properties than any of the derivatized amino acids and thus does not disturb the identification and quantitation of amino acid analysis. The reaction is simple and occurs in a matter of seconds. The amino acids can be injected on the HPLC without futher sample preparation.

15.2.2
Postcolumn Derivatization

Classical amino acid analysis was introduced as postcolumn derivatization. The amino acids were separated on a sulfonated cation-exchange column. This method offers the highest reproducibility and accuracy. The combination of postcolumn derivatization and separation of amino acids on HPLC ion exchange column provides the high speed and sensitivity necessary for quantitative routine analysis.The amino acids are eluted from the HPLC column with a continuous gradient system, from pH 3 to pH 10. After separation, the amino acids are derivatized on-

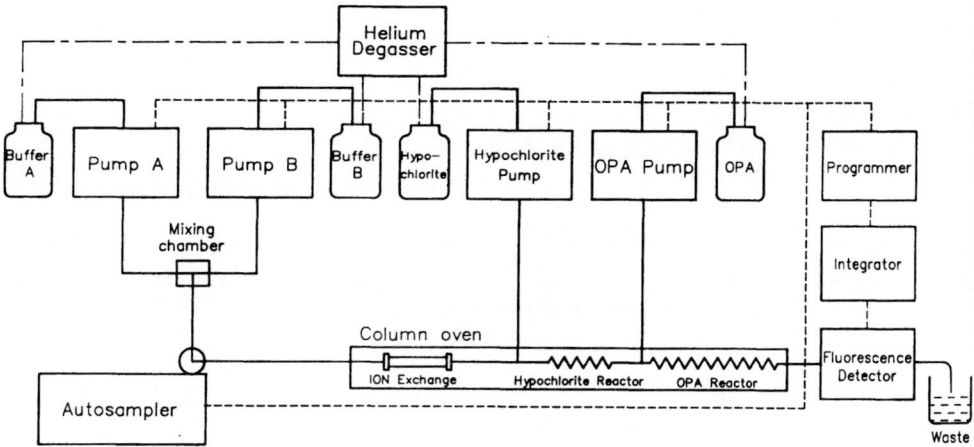

Fig. 2. Amino acid analyzer with postcolumn derivatization

stream with OPA, which allows detection of primary acids only, hence an additional reaction with hypochlorite is performed.

Hypochlorite is delivered only when proline or hydroxyproline reaches the reactor, otherwise the fluorescence of all primary amino acids would be reduced drastically. It reacts with imino acids and forms a substance accessible to the fluorescent reagent OPA. The fluorescence response for proline amounts to 50%–60% compared with the mean concentration of all primary acids (Kamp 1990). The scheme of the amino acid analyzer with postcolumn derivatization is shown in Fig. 2.

15.3
Materials and Methods

- HPLC equipment: 2 pumps M6000A, Millenium software (both from Waters, Milford MA); fluorescence detector (Shimadzu, Duisburg, Germany), autosampler Promis II (Spark, Ve Emmen, Holland)

HPLC system

15.3.1
OPA Precolumn Derivatization

- column: reversed-phase C18 Hypersil (250 × 4 mm; Knauer, Berlin, Germany)

Reagents and buffers

- OPA reagent: 100 μg ortho-phthaldialdehyde in 1000 μl methanol, 60 μl β-mercaptoethanol and 9 ml 1 M boric acid (pH 10.4)

- sample buffer: 16.9 g sodium citrate (or 19.6 g sodium citrate × 2H$_2$O) 20 ml thiodiglycol, 30 ml mercaptoethanol; bring to 1 l with water (pH 2.2)

HPLC buffers
- buffer A: 90% 12.5 mM disodium hydrogen phosphate (pH 6.5) and 10% methanol

- buffer B: methanol and 3% (v/v) tetrahydrofuran

Derivatization reaction
For good reproducibility use a fully automated derivatization system only, e.g., autosampler Promis II. Use only fresh prepared OPA solution for derivatization of amino acid. This solution should be prepared weekly and stored at –20 °C.

1. Amino acid standards or analyzed samples are dilute in sample buffer to a concentration of 100 pmol/10 µl.

2. The amino acid reaction is performed automatically using the Promis autosampler at ambient temperature for 1 min. The reagent: sample (v/v) ratio is 1:1 (10 µl each).

3. The separation (gradient) conditions are as follows:15% B to 20%B in min, 20%B to 32%B in 12 min, 32%B to 60%B in 10 min, 60%B to 90%B in 3 min, 90%B to 15%B in 2 min; 15%B corresponds to 85%A, 20%B corresponds to 80%A.

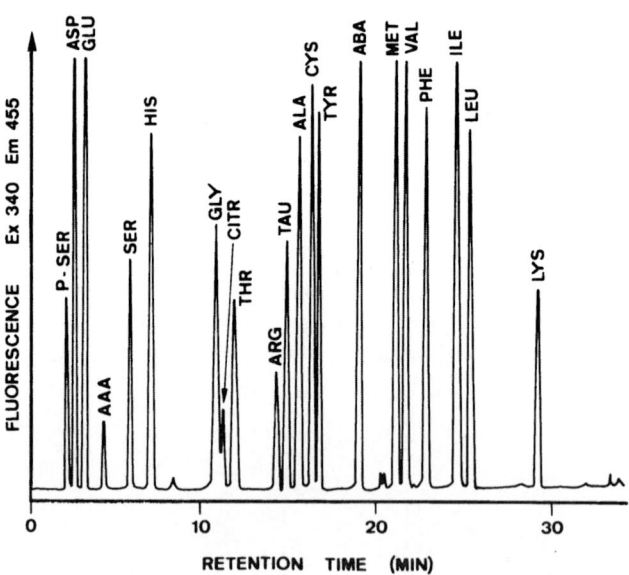

Fig. 3. Separation of 50 pmol physiological amino acid standard on reversed phase C18 column after OPA precolumn derivatization. For details, see text (*p-Ser*, phosphoserine; *Citr*, citrulline; *AAA*, amino adypic acid; *ABA*, amino butyric acid)

The flow rate is 1.0 ml/min and the column oven temperature is 50 °C. Fluorescence detection is: excitation 340 nm, emission 455 nm.

Figure 3 shows the separation of physiological amino acid standards under the following conditions:

4. Inject 100 pmol amino acid standard in 10 µl sample buffer and restart gradient with sample.

5. Calibrate the HPLC system.

6. Inject about 20 pmol protein hydrolyzate and start gradient.

7. Calculate amino acid composition.

15.3.2
NBD-Cl Precolumn Derivatization

– HPLC system: same as for OPA precolumn derivatization (see above)	**Materials**

– reversed phase C18 column, 250×4 mm (Hypersil)

– 5 mM/ml NBD-Cl in methanol

– 7.5 µM/ml OPA in 0.1 M sodium borate (pH 10.4)

– HPLC buffer A: 78% 50 mM sodium acetate (pH 7.0) in methanol

– HPLC buffer B: 25% 200 mM sodium acetate (pH 7.0) in methanol

1. The amino acid standard or sample is diluted with 0.1 M sodium borate (pH 10.4) to a concentration of 100 pmol/10 µl in an Eppendorf tube. Use 10 µl of sample for derivatization procedure.   **Derivatization procedure**

2. Pipette 10 µl of 5 mM/ml NBD-Cl in methanol to standard or sample

3. Add 10 µl of 7.5 µM/ml OPA in 0.1 M sodium borate (pH 10.4) (without mercaptoethanol).

4. Incubate the sample in a thermoblock or oven at 60 °C for 5 min.

5. Test the HPLC column (reversed phase C18, 250 × 4 mm) efficiency by injection of 100 pmol amino acid standard.

6. Calibrate the HPLC system

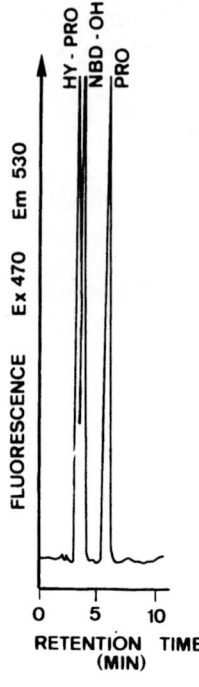

Fig. 4. Separation of 100 pmol amino acid standard on Hypersil C18 column (250 × 4 mm) after derivatization with NBD-Cl. For details, see text

7. After incubation (step 4) and cooling, inject the sample on the HPLC column.

Separation conditions

8. Separate derivatized amino acids under the following conditions:

   - fluorescence detector: excitation 470 nm, emission 530 nm

   - flow rate: 0.3 ml/min

   - gradient: 0%B to 50% B in 10 min, 50% B to 0% B in 2 min

   - column oven: 30 °C

     Figure 4 shows the separation of 100 pmol amino acid standard under the above conditions.

9. Calculate amino acid composition.

15.3.3
FMOC Derivatization

Materials

   - HPLC system (see above)

   - reversed phase column C18, 250 × 4 mm (see above)

- 2.5 mM FMOC-Cl solution in acetonitrile
- 10 mM ADAM (1-adamantanamine) in 0.5 mM boric acid (pH 7.7)
- pentane
- HPLC buffer A: 80% 50 mM sodium acetate (pH 4.1) in acetonitrile
- HPLC buffer B: 30% 50 mM sodium acetate (pH 4.2) in acetonitrile

1. Dissolve dry sample (about 100 pmol) in 100 µl 0.5 mM boric acid (pH 7.7). **Derivatization procedure**

2. Dissolve FMOC-Cl in acetonitrile to concentration of 2.5 mM

3. Pipette 100 µl FMOC-Cl solution into 100 µl sample solution and mix carefully.

4. Let reaction stand for 1 min at ambient temperature

5. Remove excess reagent using n-pentane extraction or ADAM-reaction.

5a. For extraction, use 1.5 ml n-pentane after careful mixing. After 10 min of extraction, remove the upper phase. For the second extraction again use 1.5 ml n-pentane; mix carefully and allow to separate for 10 min. Remove the upper phase containing excess reagent. The amino acid in the lower phase can be used for futher amino acid analysis.

5b. For the ADAM-reaction, prepare 10 mM ADAM solution in 0.5 mM boric acid (pH 7.7). Pipette 100 µl 10 mM ADAM into FMOC-derivatized sample. Let reaction stand for 10 min at 25 °C.

6. Separate derivatized amino acids on HPLC system.

7. Use the following conditions for separation of amino acids: **Separation conditions**
   - Fluorescence detection: excitation 260 nm, emission 310 nm
   - Flow rate: 1 ml/min
   - Column oven: 50 °C
   - Gradient: 0%B to 50% B in 40 min, 50%B to 90% B in 10 min, 90%B to 0% B in 2 min.

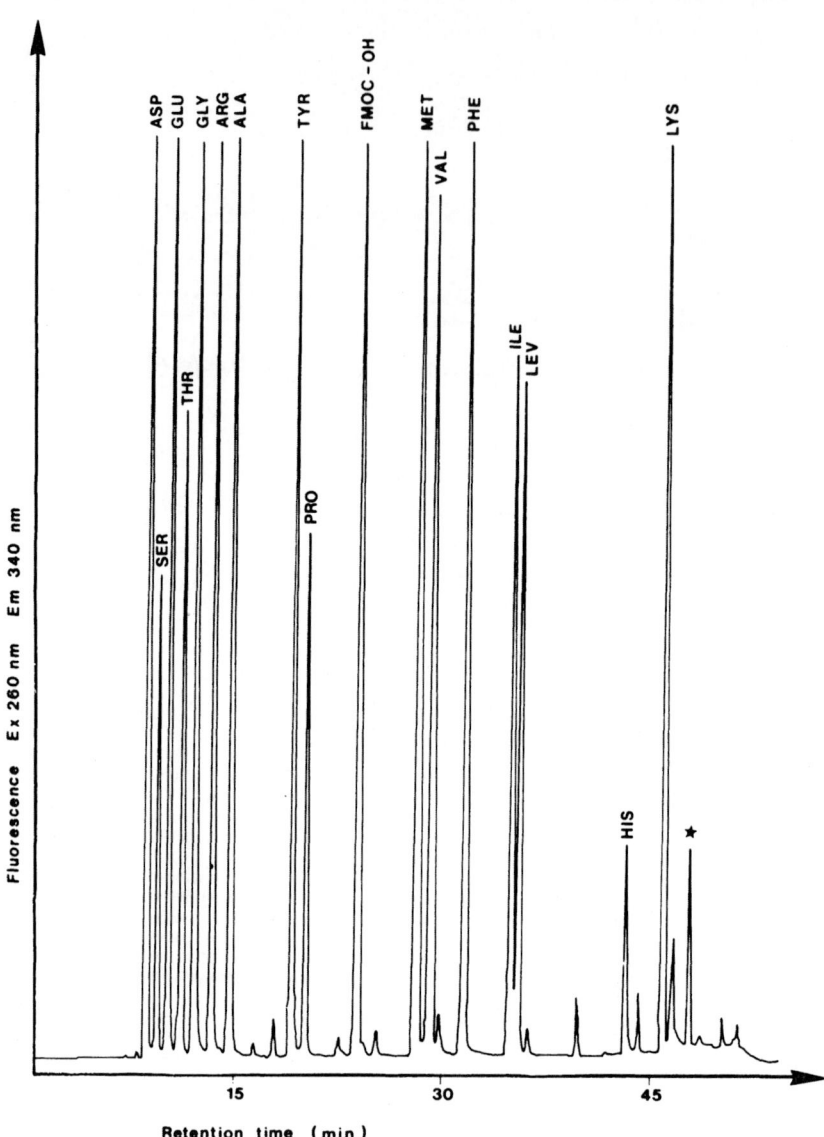

Fig. 5. Separation of 50 pmol amino acid standard after precolumn derivatization with FMOC and extraction with n-pentane. For details, see text

- The separation of amino acids under the above conditions is shown in Fig. 5.

8. Equilibrate the HPLC column for about 15 min.

9. Start above gradient without injection of sample.

10. When the baseline is stable, inject 100 pmol of FMOC-Cl derivatized amino acid standard.

11. Calibrate the HPLC system.

12. Inject 50 µl of sample.

13. After amino acid analysis, calculate amino acid composition

15.3.4
Dabsyl-Cl Precolumn Derivatization

- HPLC (see above)   **Materials**
- C18 reversed phase column: Spherisorb ODS I, 5 µm particle size, 250 × 4 mm (Knauer, Berlin, Germany)
- thermoblock
- twice-recrystallized dabsyl-Cl: prepared freshly in acetonitrile at a concentration of 2 nmole/µl
- 50 mM sodium bicarbonate (pH 8.1)
- HPLC buffer A: 25 mM sodium acetate (pH 6.5)
- HPLC buffer B: acetonitrile

1. Immediately before derivatization, dilute the lyophilized   **Derivatization** sample with 50 mM sodium bicarbonate (pH 8.1) to a concen-   **procedure** tration of 100 pmol/10µl.

2. Pipette 10 µl dabsyl solution into 10 µl sample solution.

3. Perform the derivatization reaction at 60 °C for 10 min.

4. Separate derivatized amino acids by HPLC. Use the following   **HPLC separation** conditions:

   - Flow rate: 1 ml/min
   - UV/VIS detector: 436 nm
   - Column temperature: 50 °C
   - Gradient: 15% to 25% B in 10 min, 25% B to 40% B in 20 min, 40% B to 70% B in 7 min. Hold at 70% B for 2 min, then 70% B to 15% B in 2 min.

   The separation of amino acids under the above described conditions is shown in Fig. 6.

5. Start gradient without injection of sample.

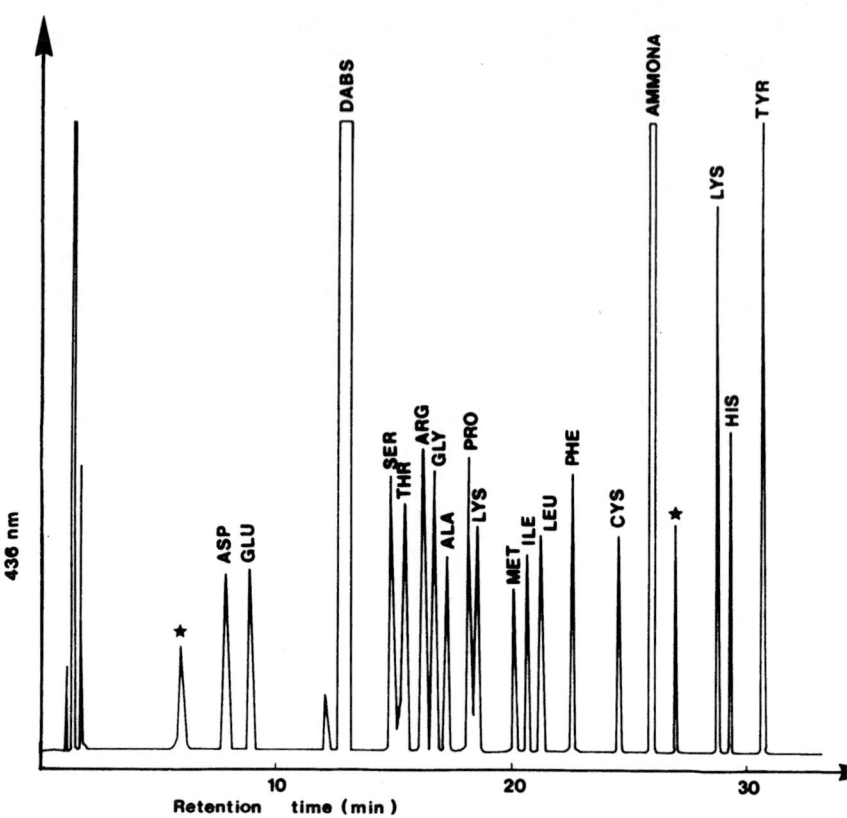

Fig. 6. Separation of 100 pmol amino acid standard mixture and precolumn derivatization with dabsyl chloride. For details, see text

6. When the baseline is stable, inject 100 pmol of derivatized protein standard and start gradient again.

7. Calibrate the HPLC system.

8. Inject 10 µl of the derivatized sample.

9. Calculate amino acid composition.

15.3.5
OPA Postcolumn Derivatization

HPLC system   The postcolumn OPA amino acid analyzer consist of two HPLC pumps, a dynamic mixing chamber, a fluorescence detector, a column compartment with temperature control unit, an auto-sampler and a reagent delivery system with two pumps and two reactors. Separate the sample on a short 120×4 mm resin cation-exchange column (e.g., Eurokat, Knauer Scientific Instruments, Berlin, Germany) covered with sodium or lithium ions for good resolution of protein hydrolyzates or physiological samples. Use the following conditions:

- Column temperature: 68 °C
- Flow rate: 0.4 ml/min
- Fluorescence detection: excitation 345 nm, emission 455 nm
- starting buffer A: 0.2 mM sodium citrate (pH 3.0)   **HPLC buffers**
- buffer B: 0.2 mM sodium borate (pH 10.0)
  Degass all buffers for 15 min before use using an ultrasonic bath or continuously using, e.g., the ERMA degasser (ERC, Alteglofsheim, Germany).

- gradient: 3%B to 5%B in 10 min, 5%B to 10%B in 6 min, 10%B to 15%B in 8 min, 15%B to 25%B in 3 min, 25%B to 30%B in 5 min, 30%B to 50%B in 5 min. Hold at 50%B for 5 min, then 50%B to 80%B for 5 min. Hold at 80%B for 5 min, then 80%B to 3%B for 5 min.
  The separation of amino acids under the above described conditions is shown in Fig. 7.

After separation, the amino acids are derivatized on-stream with   **Reagents**
OPA. Hypochlorite is delivered only when proline or hydroxyproline reaches the reactor.

- OPA reagent: 6 mM OPA /12 mM mercaptopropionic acid (MPA) reagent. Prepare in 1 l 0.5 M potassium borate buffer (pH 10.4).
- hypochlorite reagent: Add 400 µl hypochlorite (e.g., from Aldrich, 5% free chloride) to 100 ml 0.5 M potassium borate buffer.
  Degass both reagents for 20 min using ultrasonic bath.

- For derivatization of all amino acids with OPA/MPA, use 6m   **Reactors**
  knitted teflon tube (diameter 0.25 mm).
- For derivatization of secondary amino acids use 2m knitted teflon tube (diameter 0.25 mm).

1. Heat the OPA reactor in a water bath or column compartment   **Derivatization**
   at 65 °C. Use a flow rate of 0.4 ml/min for detection of secon-   **procedure**
   dary amino acids and 0.2–0.1ml/min for detection of all primary amino acids

2. Heat the hypochlorite reactor at 65 °C and swich on during the proline elution only. Use 0.1 ml flow rate.

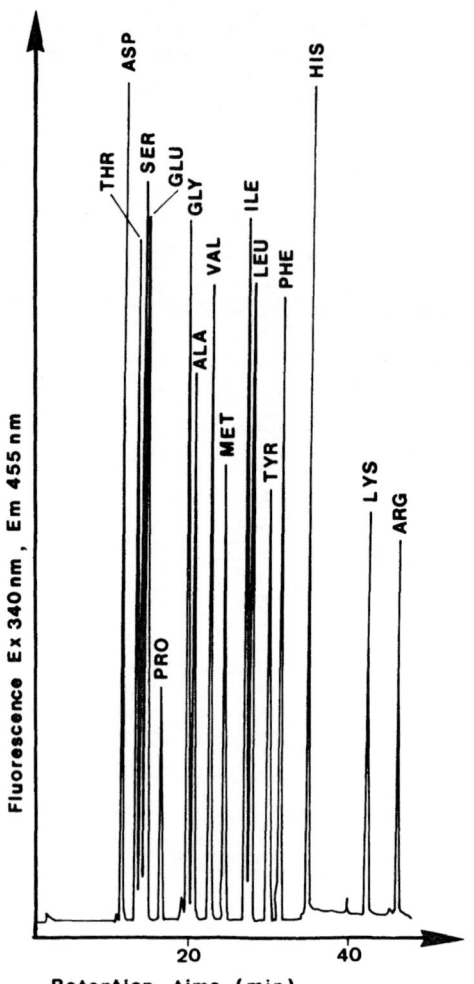

Fig. 7. Separation of 50 pmol amino acid standard after postcolumn derivatization using OPA and hypochlorite reagents. For details, see text

Note: Hypochlorite destroys all primary amino acids. Be sure to switch off the delivery pump when the primary amino acids reach the reactor.

Calibration and analysis

1. Start gradient without injecting any sample.

2. When the baseline is stable, inject 100 pmol amino acid standard in 20 µl buffer A.

3. Calibrate amino acid analyzer using amino acid standard as external standard.

4. Add 50 µl buffer A to lyophilized protein hydrolyzate or physiological sample.

5. Separate the sample using the same gradient given for separation of amino acid standard.

6. Calculate amino acid composition.

The on-line postcolumn derivatization, with hypochlorite and the fluorescence reagent OPA, eliminates manual preparation steps. Moreover, it provides optimal reproducibility and sensitivity at the low picomole level. The analyzer can be modified to perform precolumn derivatization and sample purification.   **Comments**

15.4
Calculation of Amino Acid Composition

1. Calibrate amino acid analyzer with 100 pmol standard amino acid mixture. The standard amino acids are separated and the peak areas of each peak correspond to 100 pmol.

2. The amounts for the various amino acids in the analyzed sample are compared with the quantity of the amino acid in the standard mixture (e.g., 100 pmol)

3. The quantity of each amino acid is calculated from the following equation:

$$\frac{\text{Area of amino acid in the sample}}{\text{area of amino acid in the standard mixture}} \times 100 = X \text{ (pmol)}$$

if the standard mixture contains 100 pmol each amino acid.

4. If the molecular weight of the analyzed protein is unknown, only Mol% of the analyzed sample can be determined. Mol% will be calculated as:

$$\frac{\text{pmol of analyzed amino acid x molecular mass}}{\text{sum of weights (mol x molecular mass) found for all amino acids}} \times 100 = \text{Mol}\%$$

5. If the molecular weight of the analyzed protein is known, then the amino acid composition can be calculated as follows:

   – The weight of each amino acid will be determined after multiplying Mol% by the molecular mass of the protein.

The number of residues will be determined after dividing by the molecular weight of the amino acid.

- The amount of protein can be determined after dividing the sum of the pmole amounts of all amino acids by the sum of all amino acids residues. Determination of the concentration of the analyzed protein is possible only when the molecular mass is known.

15.5
Advantage of Amino Acid Analysis

The amino acid analysis allows determination of the following:

- The molar percentage of individual amino acids.
- The amino acid composition (the number of amino acid residues per mole), if molecular weight is known.
- The concentration of protein in, e.g., pmoles.
- The purity of the protein. When the number of all amino acids are whole the sample is pure, in all other cases the sample is contaminated.
- Homology studies, using protein data bank.

15.6
Maintenance of Columns

1. Use column in connection with a precolumn (2–4 mm) packed with the same material as in the separation column.

2. Prior to first use, wash column with 80% methanol in water to remove all impurities and then with the aqueous buffer for 2 h. Rinse overnight, by running a gradient of 0%–80% buffer B in 2 h with the buffers to be used the next day. Keep the flow rate at 0.5 ml/min.

3. Check the column purity by a blank gradient without sample injection.

4. Test the column efficiency by running an amino acid standard.

5. Inject sample solute in sample buffer for OPA derivatization or derivatized sample after FMOC, dabsyl, or NBDCl derivatization.

6. In case of "ghost" peaks appearing in chromatogram, regenerate the column as follows:

   - Wash column with water to obtain stable baseline.
   - Inject 4 × 200 μl DMSO (dimethylsulfoxide).
   - Wash column with 100 ml water.
   - Replace water by methanol and wash the column to obtain a stable baseline (about 100 ml).
   - Replace methanol by chloroform and wash out all impurities.
   - Repeat column wash with methanol (100 ml) to obtain a stable baseline.
   - Equilibrate column with starting aqueous buffer.
   - Check the resolution by injection of amino acid standard mixture.

7. Store column after washing with methanol.

15.7
Results and Comments

Amino acid analysis by HPLC is a quick and sensitive method. The choice of suitable reagents depends on both the area of application and the nature of the problem. Two parameters are decisive – sensitivity and speed. The highest sensitivity can be achieved using fluorescence reagents such as OPA and FMOC, although they are limited in their determination of secondary amines and tryptophan, respectively. Dabsyl chloride allows determination of all amino acids, including cysteine and proline; however, sensitivity of the analysis is lower and the life of the column is shorter. High sensitiviy is important in the research laboratory, where sample quantities are limited and repeated isolation of new biological material can take months. New sensitive methods of analysis must therefore be applied. A fast and fully automated method is important for clinical applications and in the food and animal feed industries. For a routine method, the practical aspects are more important. Derivatization with OPA is the simplest technique, as it requires a minimum of manipulation steps and can be fully automated. Specifity, linearity, and precision are also important criteria for all routine methods. Derivatization with OPA is the most specific method. All other reagents cause the formation of by-products that impair separation and identification of amino acids of interest. Conditions for all reactions must be optimized to provide a reliable linear response. Accuracy depends on the method of quantitation and the stability of retention times. It is important to carefully calibrate the equipment and choose the integration parameters so that the various peaks on the chromatogram can be correctly identified. The greatest precision of a derivatization reaction can be achieved only in a fully automated system. The on-line postcolumn derivatization with hypochlorite and the fluorescence reagent OPA eliminates manual preparation steps and provides optimal reproducibility and sensitivity at the low picomole level. The combined two-step reaction with hypochlorite and OPA allows determination of primary and secondary amino acids. The combination of HPLC and postcolumn derivatization provides the high reproducibility, accuracy, speed and sensitivity necessary for quantitative routine analysis.

Even though technical developments in new amino acid analyses have attained the femtomole level, contaminants in the analyzed samples still limit highly sensitive analyses. In the future, it is important to develop methods for sample preparation to minimize sample background. Only pure, salt-free sample guarantees reliable amino acid analysis at the picomole level (as otherwise the impurities are analyzed and not the sample).

References

Carisano A (1985) Rapid and sensitive method for the determination of proline by reversed -phase high-performance liquid chromatography with automated precolumn fluorescence derivatization. J Chrom 318: 132–138

Chang J.-Y, Knecht R, Braun DG (1983) Amino acid analysis in the picomole range by precolumn derivatization and high-performance liquid chromatography. Methods Enzymol 91: 41–49

Einarsson S, Josefsson B, Lagerkvist S (1983) Determination of amino acids with 9-fluorenylmethylchloroformate and reversed-phase HPLC. J Chromat 282: 609–618

Kamp RM, Vollenbroich D (1990) HPLC-Aminosäureanalytik mit Nachsäulenderivatisierung. BioTech. Nov. 38–41

Palmerini CA, Fini C, Floridi A, Morelli A, Vedovelli A (1985) HPLC analysis of free hydroxyproline and proline in blood plasma and of free and peptid bound hydroxyproline in urine. J Chromat 339: 285–292

Roth M (1972) Fluorescence reaction of amino acids. Anal Chem 43: 880–882

Spackmann DH, Stein W, Moore S (1958) Automatic recording apparatus for use in the chromatography of amino acids. Anal Chem 30: 1190–1206

Chapter 16

Quantitative Analysis of D- and L-Amino Acids by HPLC

D. VOLLENBROICH and K. KRAUSE

16.1
Introduction

In proteins only the L-amino acid enantiomeric form appears, but in most organisms D-amino acids are also present either as structural elements of peptides or in a free form. The enzymatic synthesis of D-amino acids is widespread in microorganisms taking place during the assembly of the bacterial cell wall or the biosynthesis of secondary metabolites, such as antibiotics. From these sources D-amino acids find their way into animals, plants, food and beverages. Therefore, determination of the D-amino acid concentration is relevant in the fields of peptide research, medicine, and food technology.

For the separation of L-amino acids precolumn derivatization combined with high performance liquid chromatography (HPLC) is one of the most popular methods. This technique can also be used for the separation and quantitative analysis of D- and L-amino acids. The derivatization of all primary D- and L-amino acids can be achieved with o–phthaldialdehyde (OPA) and a thiol (Roth 1972) increasing the sensitivity by fluorimetric detection to the femtomole level. This derivatization technique can easily be performed, with the additional advantages of very short reaction times and high derivatization yields (for detailed information see Chap. 15). For high chromatographic resolution of the optical isomers, introduction of an asymmetric environment is necessary (Manning and Moore 1968; Hare and Gil-Av 1979). This is achieved intramolecularly by conversion of each amino acid to a diastereomer using OPA together with N-isobutyryl-L-cysteine or N-isobutyryl-D-cysteine as the thiol and forming a fluorescent dipeptidic isoindole (Brückner 1991):

OPA   amino acid   N-isobutyryl-L-cysteine

Here, we describe a fast and easy reversed phase HPLC method that separates nearly all proteinogenic L-amino acids from their D–enantiomers with high resolution in a single chromatographic run and a detection limit in the picomole level (Brückner 1991, 1992). This method enables a fully automated analysis resulting in higher reproducibilities than obtained with manual procedures. The disadvantage is that the secondary amino acids proline and hydroxyproline cannot directly be derivatized by OPA. Thus, an additional oxidation step before derivatization and a separate chromatographic run for the analysis of these secondary amino acids is necessary. Also, the OPA/N-isobutyryl-cysteine system is not suitable for the determination of cyst(e)ine, resulting in interfering peaks of low fluorescence intensity (Brückner 1991). To overcome this problem, methods involving the use of chiral stationary phases (Merino 1990), chiral eluants (Hare and Gil-Av 1979), or other diastereomeric dipeptides (Manning and Moore 1968) are described. An overview has been given by Lindner (1988).

16.2
Materials

Equipment   – HPLC gradient system including: fluorescence detector, integrator, column heater, degasser

– optional: autosampler with derivatization function

– HPLC column, reversed phase C_{18}, 250 mm × 4.6 mm i.d., 4 µm, 60 Å, e.g., Nova-Pak (Waters, Milford, MA, USA)

– optional: guard column, reversed phase C_{18}, 10 mm × 3.9 mm i.d., 4 µm, 60 Å, e.g., Nova-Pak (Waters, Milford, MA, USA)

– OPA/N-isobutyryl-L-cysteine chiral derivatization reagent (e.g., Pierce, Rockford, IL, USA; or GROM, Herrenberg, Germany). A procedure for the preparation of the reagent N-isobutyryl-L-cysteine is described by Brückner (1991).

- D-, L-amino acid standards (e.g., Sigma, Deisenhofen, Germany; or GROM, Herrenberg, Germany)

- internal standard, e.g., L-homo-arginine (ICN, Meckenheim, Germany) or L-norvaline (Sigma, Deisenhofen, Germany)

- Derivatization buffer: 1 M borate buffer (pH 10.7): Weigh out **Buffers** 61.8 g boric acid and dilute in approx. 800 ml deionized water. Adjust the pH with KOH. Bring final volume to 1 l with deionized water. The detection limit is mainly effected by the quality of the boric acid used in the derivatization buffer. For sensitive analysis use boric acid of ultra grade quality, e.g., Suprapur (Merck, Darmstadt, Germany). This buffer should be stored at room temperature to prevent crystallization.

- OPA reagent: Dilute the OPA/N-isobutyryl-L-cysteine derivatization reagent to a concentration of 260 mM N-isobutyryl-L-cysteine and 170 mM OPA with derivatization buffer directly before use, or follow the instructions of the manufacturer. Store the OPA reagent at –20 °C not longer than 2 days. It is advisable to always freshly prepare the solution before use.

- eluent A: 23 mM sodium acetate buffer (pH 6.0): Weigh out 1.88 g sodium acetate and dilute in approx. 800 ml deionized water. Adjust the pH with acetic acid. Bring final volume to 1 l with deionized water.

- eluent B: 600 ml methanol mixed with 50 ml acetonitrile

The eluents should be filtered (0.45 μm filter) and degassed either with a water pump, by sonification, or automatically, e.g., by an ERMA (Japan, distributed by ERC, Alteglofsheim, Germany) degasser. Acetonitrile is a carcinogen if inhaled or in contact with skin. Avoid any contact, use a fume hood during eluent preparation and close the eluent and the waste bottle tightly. Properly dispose of waste generated by use of this toxic substance.

16.3
Experimental Procedures

The amino acids must exist in free form, which is achieved by **Sample** extraction with ethanol and after total hydrolysis of proteins or **preparation** peptides. The total amount of each amino acid in the sample should not be higher than 5 nmol to keep within the linear range of the peak area/concentration ratio. The sample should be free of salt.

1. Fill the sample solution into a reaction tube.

2. Remove the solvent in vacuo.

3. Dilute the sample in 0.1 N HCl to approx. 200 pmol/µl.

4. The amino acid standard solution should contain glycine, the D- and L-amino acids and L-norvaline or L-homo–arginine as the internal standard at a concentration of 200 pmol/µl each in 0.1 N HCl.

Derivatization

1. Transfer 25 µl of the amino acid solution into a fresh reaction tube.

2. Mix the solution and the OPA reagent in a ratio of 1:1. Close the tube, vortex carefully and incubate the reaction mixture for 90 s at 25 °C.

3. Inject 10 µl of the sample immediately after derivatization onto the HPLC system. OPA derivatives are only stable for approx. 2 min. A derivatization device as part of an auto-sampler, the Promis II from Spark Holland (Ve Emen, Nether-lands) for example, can improve the reproducibility. Such fully automated systems mix the sample and the derivatiza-tion reagent in the sample loop and the injection needle for a programmable reaction time before injection.

Chromatogra-phic procedure and evaluation

1. The flow rate of the eluent is 1 ml/min.

2. The fluorescence detector should be set at $\lambda = 300$ nm for the excitation and $\lambda = 445$ nm for the emission.

3. Separate the amino acids in a linear gradient from 0% eluent B to 53.5% eluent B in 75 min. Wash the column with 100% eluent B for 10 min and equilibrate with 100% eluent A for 5 min.

4. Keep the column at a temperature of 30°C.

5. The concentrations of the amino acids are calculated compar-ing the areas of the amino acid peaks of the sample and the corresponding amino acid from the standard prepared under the same conditions or by correlation to the internal stan-dard. A detailed description is presented in Chap. 15. The in-tensity of the fluorescence signal depends on the enanti-omeric form. In this method the diastereomers obtained from the L-amino acids usually show a reduced fluorescence signal compared to the D-enantiomers. Therefore the ratio of the enantiomers should be quantified using the fluorescence fac-tor F:

$$F = \frac{\text{peak area D- amino acid}}{\text{peak area L- amino acid}}$$

6. Calculate the fluorescence factor F from the fluorescence signals of a racemic amino acid mixture as an average of several measurements.

7. Multiply all fluorescence signals received from L-amino acids with the corresponding F factor before calculating the ratio of the enantiomers. It is not necessary to perform this calculation if the ratio can be calculated from concentration values obtained from an integrated standard amino acid chromatogram.

amino acid	Asp	Glu	Ser	His	Ala	Arg	Tyr	Val	Met	Phe	Leu	Ile	Lys
fluorescence factor (F)	0.950	1.117	1.078	1.070	1.117	1.140	1.116	0.896	1.234	0.881	0.905	0.955	1.085

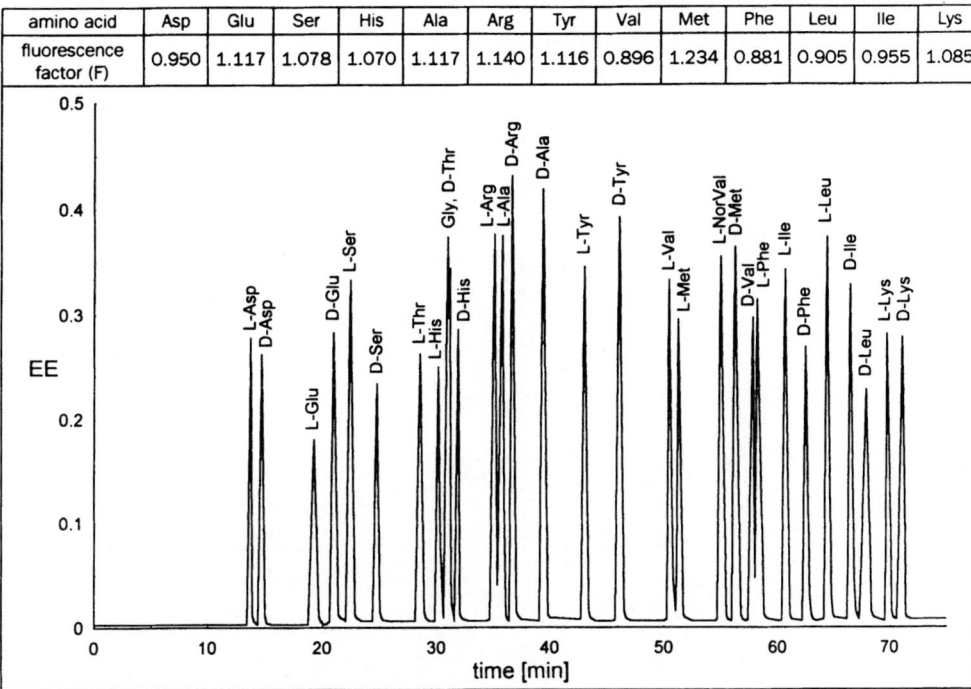

Fig 1. Separation of a D-, L-amino acid standard after derivatization with o-phthaldialdehyde and N-isobutyryl-L-cysteine. A mixture of 14 pairs of amino acid enantiomers (1 nmol of each of the L- and D-forms in 10 µl) was injected. The amino acids were eluted with a linear gradient from 0% to 53.5% methanol/acetonitrile (600 ml + 50 ml) in a sodium acetate buffer (23 mM, pH 6.0) in 75 min. The internal standard was L-norvaline (1 nmol)

16.4
Results and Comments

Figure 1 shows the elution profile of the complete separation of nearly all proteinogenic L-amino acids from their corresponding D–enantiomers and the achiral glycine derivatized with the OPA/chiral thiol technique. All diastereomeric amino acid derivatives are almost baseline resolved with a simple linear gradient. The enantiomers of physiological amino acids can be separated in the same way. 100 pmol of the isoindole derivatives are routinely quantifiable by peak area integration of the fluorescence signal with high precision. The reproducibility of the fully automated system was higher than 99.8%. These facts demonstrate that the described method is suitable for the detection and quantification of free D- and L-amino acids in a broad range of applications. An example is shown in Fig. 2, in which the amino acid composition of the microbially synthesized lipopeptide antibiotic surfactin was investigated. The fluorescence factors F (Fig. 1) calculated from the standard chromatogram demonstrate the inequalities of the fluorescence signals derived from the L– and D–enantiomers. These factors are therefore necessary for correct calculation of the amino acid composition in samples such as the composition of the peptide portion of surfactin.

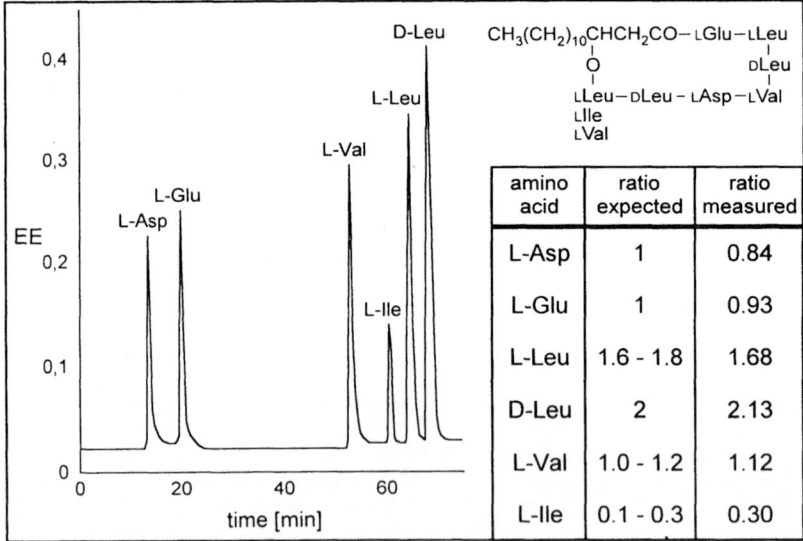

amino acid	ratio expected	ratio measured
L-Asp	1	0.84
L-Glu	1	0.93
L-Leu	1.6 - 1.8	1.68
D-Leu	2	2.13
L-Val	1.0 - 1.2	1.12
L-Ile	0.1 - 0.3	0.30

Fig 2. Separation of a surfactin hydrolysate, a lipoheptapeptide antibiotic from *Bacillus subtilis*. All separation conditions were as described in Fig. 1. Approx. 1 nmol of the hydrolyzed surfactin was injected. The ratio of the enantiomers was calculated based on the fluorescence factors F presented in Fig. 1

16.5
Troubleshooting

Temperature has a significant effect on peak resolution. The general rule "the higher the temperature, the higher the resolution" is still effective in this case. Be aware that if higher temperatures are used during chromatography, racemization may occur. In some cases it may be of advantage to reverse the elution order of the D- and L–enantiomer pair of each amino acid. This can be easily performed using N-isobutyryl-D-cysteine with OPA for derivatization (Brückner 1991). Add 0.01% (w/v) sodium azide (NaN$_3$) or commercially available buffer preservation solutions to eluent A to prevent microbial contaminations.

References

Brückner H, Jaek P, Langer M, Godel H (1992) Liquid chromatographic determination of D-amino acids in cheese and cow milk. Implication of starter cultures, amino acid racemases, and rumen microorganisms on formation, and nutritional considerations. Amino Acids 2: 271–284

Brückner H, Wittner R, Godel H (1991) Fully automated high–performance liquid chromatograpic separation of DL-amino acids derivatized with o-phthaldialdehyde together with N–isobutyryl–cysteine. Application to food samples. Chromatographia 32: 383–388

Hare PE, Gil–Av E (1979) Separation of D and L amino acids by liquid chromatography: Use of chiral eluants. Science 204: 1226–1228

Lindner W (1988) Indirect separation of enantiomers by liquid chromatography. In: Zief M, Crane LJ (eds) Chromatographic chiral separations. Dekker, New York, pp 91–130

Manning JM, Moore S (1968) Determination of D- and L-amino acids by ion exchange chromatography as L–D and L–L dipeptides. J Biol Chem 243: 5591–5597

Merino IM, Gonzalez EB, Sanz-Medel A (1990) Liquid chromatographic enantiomeric resolution of amino acids with β–cyclodextrin bonded phases and derivatization with o–phthalaldehyde. Anal Chim Acta 234: 127–131

Roth M (1972) Fluorescence reaction for amino acids. Anal Chem 43: 880–882

Part VI
Identification of Cysteine Residues and Lipids in Proteins

Identification of Cysteine Residues and Disulfide Bonds in Proteins

T. A. Egorov

17.1
Introduction

The sulfhydryl groups of cysteine residues are the most reactive among the protein amino acid residues of a protein, participating in reactions such as addition, substitution, alkylation, oxidation, and thiol-disulfide exchange. Sulfhydryl groups react rapidly with heavy metals and undergo decomposition by irradiation. They also play an important role in the biological activity of proteins, such as in the cysteine class of enzymes. This very reactive group is highly susceptible to oxidation to the disulfide by atmospheric oxygen in the presence of iron salts or by other mild oxidizing agents. The oxidation product is cystine, in which the disulfide bond constitutes a covalent bridge between two residues of cysteine. Disulfide cross-links can be cleaved by reducing agents, e.g. mercaptoethanol or dithiothreitol, yielding two molecules of cysteine from cystine.

Here, two protocols for identification of cysteine residues and disulfide bonds in proteins will be outlined. The advantages of these selective isolation methods are:

- Due to the selectivity of this method, cysteine-containing peptides may be effectively isolated and identified from large and complex proteins.

- Even very hydrophobic proteins, such as membrane proteins, are easily digested with the enzymes.

17.2
Principle and Applications

17.2.1
Identification of Cysteine Residues in Proteins by Covalent Chromatography

Several methods for the identification of cysteine residues in proteins based on selective labeling of these residues have been proposed; however, none of them is effective enough, especially for large and/or membrane proteins. A selective method for isolation of cysteine-containing peptides by covalent chromatography has been proposed (Egorov et al. 1975). The method is based on the covalent attachment of proteins at free SH-groups to an insoluble activated support (thiopropyl Sepharose 6B, Fig. 1) by means of a thiol-disulfide exchange reaction with subsequent enzymatic digestion of immobilized protein (conjugate) (Fig. 2). Multiple attachment will occur if the protein has several cysteine residues. The cysteine-containing peptides formed will then become attached to the support. After digestion, non-cysteine-containing peptides (fraction I) are removed by washing the gel, while cysteine-containing peptides are detached from the support by reduction (fraction II). Fraction II, consisting of cysteine-containing peptides, excess reducing agent, 2-thiopyridone, formed and salts may be desalted and/or separated by reversed phase HPLC (RP-HPLC). Note: small hydrophilic peptides may be lost at this stage. Cysteine-containing peptides may be alkylated after the desalting step only, as large excesses of thiopyridone inhibit alkylation of SH-groups. Monitoring S-ethylpyri-

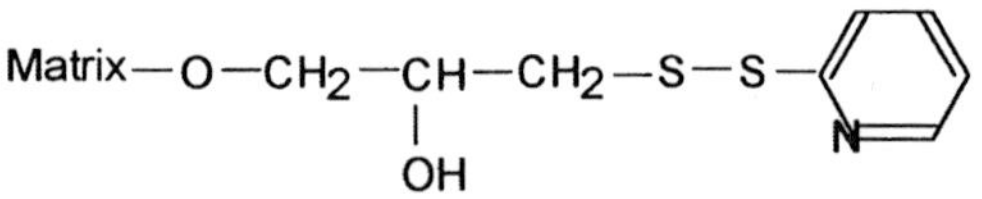

$$\text{Matrix} - \text{O} - \text{CH}_2 - \underset{\underset{\text{OH}}{|}}{\text{CH}} - \text{CH}_2 - \text{S} - \text{S} - \langle\!\!\langle \text{N} \rangle\!\!\rangle$$

Fig. 1. Partial structure of thiopropyl Sepharose 6B. The reagent is a mixed disulfide containing 2-thiopyrydyl protecting groups which are attached to Sepharose through a chemically stable ether linkage. No leakage occurs under conditions which do not damage the matrix. The 2-hydroxypropyl residue acts as a hydrophilic spacer group. Thiopropyl Sepharose 6B contains approximately 20 µmoles 2-pyridyl disulfide groups per ml gel. The protein coupling capacity varies with the protein. A typical value for ceruloplasmin (MW 132|000) is 14 mg/ml gel. Thiopropyl Sepharose 6B is supplied as a freeze-dried powder (1 g equivalent to approx. 3 ml swollen gel). Additives have been included to preserve swelling characteristics

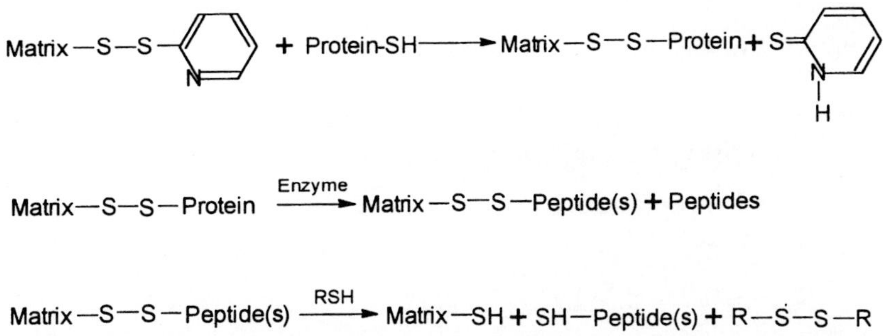

Fig. 2. Reaction scheme for covalent chromatography of thiol protein on activated thiopropyl Sepharose 6B and selective isolation of cysteine-containing peptides

dilated peptides at two wavelengths, 210–214 nm and 254 nm, may help further detection of these peptides. The procedure described below is a further modification of the above described method and is suitable for microgram level analysis of proteins.

17.2.2
Determination of Disulfide Bonds

The number of cysteine residues in proteins involved in disulfide bonding may be determined by mass analysis before and after their reduction and alkylation (Egorov et al. 1994). Protein disulfide bonds (S-S-bonds) can be localized by cleaving the intact protein between half-cystinyl residues and identifying the disulfide-containing peptides by their amino acid composition, sequences or masses. (For a comprehensive list of references and other methods of S-S-bond determination in proteins see Gray 1993.) Analysis of S-S-bonds involves the following steps:

- Alkylation of intact protein (in case of free SH-groups)
- Fragmentation of intact protein by chemical or enzymatic methods
- Separation of products by RP-HPLC at pH 2
- Reduction and alkylation of peptide fractions and their separation
- Sequence and/or amino acid analysis of alkylated peptides
- Computer/manual alignment of disulfide-bonded fragments

17.3
Isolation of Cysteine-Containing Peptides

The protocol is described for up to 1 mg of protein. The procedure consists of the following steps:

- Sample preparation
- Immobilization
- Enzymatic digestion of conjugate
- Isolation of Cys(–) peptides
- Isolation of Cys(+) peptides
- Alkylation of Cys(+) peptides
- RP-HPLC of peptides

17.3.1
Materials

Reagents
- thiopropyl Sepharose 6B: approx. 45 ml swollen gel (Pharmacia, Uppsala, Sweden), or approx. 15 ml (Sigma, St. Louis, USA)
- aldrithiol-2 (2,2'-dipyridyl disulfide, DPDS; 2,2'-dithiopyridine; Aldrich, Milwaukee, USA)
- 4-vinylpyridine (Aldrich Milwaukee, USA)
- trypsin (Promega, Madison, USA)
- guanidine hydrochloride (ultra grade; Sigma, St. Louis, USA)
- Tris base (TRIZMA, ultra grade; Sigma, St. Louis, USA)

Buffers
- buffer A (coupling/alkylating buffer): 0.5 M Tris in 6 M guanidine-HCl and 1 mM EDTA, pH 7.6
- buffer B (digestion buffer for trypsin and related enzymes): 0.1 M Tris (pH 7.6)
- buffer C (washing buffer): buffer B in 0.5 M NaCl
- buffer D (reducing buffer): buffer B in 0.5 M NaCl and 0.1 M DTT
- buffer E (storage buffer): 0.1 M sodium acetate (pH 5.5), containing 0.1% sodium azide

HPLC solvents
- solvent A: 0.1% trifluoroacetic acid (TFA)
- solvent B: 0.08% TFA in acetonitrile

17.3.2
Methods

1. Reduce protein before immobilization (only for proteins
   containing S-S-bonds): Dissolve protein in 300 µl buffer A.
   Add 10 µl DTT solution (1.4 µmol/µl) and let reaction proceed for 4 h at 37 °C or overnight at room temperature.   **Sample
   Preparation**

2. Desalt protein (for any samples) by RP-HPLC using "short"
   acetonitrile gradient, e.g., from 10%B to 60%B in 15 min, after
   washing out salts. Detect protein fraction at 214, 254 or
   280 nm and dry on SpeedVac concentrator.

Tip Column

1. Prepare tip column: Cut 1 mm from a 1 ml Gilson type tip.
   Insert small piece of glass/cotton wool as filter.

2. Attach a 2–5 ml plastic syringe with polyethylene/teflon tube
   (approx. 1 mm i.d., 10 cm lengh) to column using silicone/ty-
   gon tube as connector.

3. Calibrate tip column in 100 µl divisions.

Preparation of Thiopropyl-Sepharose 6B

1. Weight out 150 mg of thiopropyl-Sepharose 6B and place in a
   beaker/flask containing 4–5 ml water. Stir gently and leave for
   15 min with occasional stirring; decant supernatant.

2. Add 1 ml water, mix gently and transfer gel suspension with
   cut pipette into tip column. Remove excess liquid with syringe; do not dry. Avoid admitting air bubbles into the gel; if
   so resuspend gel. Transfer remaining gel into column and remove excess liquid.

3. Equilibrate gel in column with 3×500 µl buffer A. Remove
   excess buffer with syringe; avoid bubbles.

Immobilization

1. Replace tip column syringe with a 1 ml one.

2. Dissolve dried protein in 300 µl buffer A just before immobilization. Add protein into column and mix with syringe by
   carefully pulling and pushing plunger for about 500 µl; avoid
   bubbles.

3. Let stand for 4 h with ocassional mixing (every 20 min) as described in step 2.

4. Remove liquid with syringe. Wash gel with 3×100 μl buffer A. Collect combined washings if neccessary for further analysis. This fraction consists of non-bound protein(s).

5. Wash gel additionally with 5×200 μl buffer A. Discard washings.

Enzymatic Digestion of Conjugate

1. Equilibrate conjugate using 5×200 μl buffer B. Remove excess buffer.

2. Add 200 μl of enzyme solution and mix gel with syringe.

3. Seal column tightly with parafilm. Put column into thermostat and leave for 1–4 h at 37 °C.

4. Remove liquid with syringe. Wash gel using 3×200 μl buffer C. Collect combined fraction (fraction I, non-cysteine-containing peptides) if necessary for subsequent analysis.

5. Replace syringe with 2–5 ml one and additionally wash gel using 5×300 ml buffer C. Discard washings.

6. Replace syringe with a 1 ml one. Add 900 μl buffer D. Remove about 50 μl of liquid with syringe and leave for 30 min.

7. Remove liquid slowly from column and collect this fraction (fraction II, cysteine-containing peptides).

8. Wash gel with 5×300 μl of buffer C and discard washings.

Regeneration of Thiopropyl-Sepharose 6B

1. Dissolve 2 mg of 2,2'-dipyridyl disulfide (DPDS) in 2 ml of ethanol and dilute with 3 ml buffer B.

2. Regenerate gel with 6×0.5 ml DPDS solution.

3. Wash gel with 10×0.5 ml buffer B containing 0.1% sodium azide. Fill column to the top with this buffer.

4. Seal column with parafilm and keep gel at 4 °C (max. 1 month).

RP-HPLC of Cysteine-Containing Peptides

1. Equilibrate analytical RP column with 0%B or 5%-10%B for large peptides (> 15–20 residues) at flow rate of 1 ml/min and temperature of 40°–50 °C; detect at 210–214 nm.

2. Inject fraction II into column and wash column with 0%B until all salts are eluted.

3. Start gradient, e.g. from 0%B to 40%B in 60 min, at flow rate 1 ml/min and collect fractions.

4. Evaporate fractions on SpeedVac.

Alkylation of Cysteine-Containing Peptides

1. Dissolve peptide in 40 µl buffer A.

2. Add 1 µl DTT (1.4 µmol/µl) and leave to react for 10 min under nitrogen.

3. Add 2 µl 4-vinylpyridine diluted 1:10 with ethanol (approx. 2 µmol) and leave to react for 10 min.

4. Desalt and/or separate alkylated peptides immediately by RP-HPLC as described above. Detect peptides at 210–214 nm and 254 nm.

17.4
Identification of Disulfide Bonds in Proteins

The protocol is described for about 100 µg protein.

17.4.1
Materials

- cyanogen bromide (CNBr): 5 M solution in acetonitrile **Reagents** (Aldrich, Milwaukee, USA) or solid (Sigma, St. Louis, USA)

- trypsin (Promega, Madison, USA)

- 4-vinylpyridine (Aldrich, Milwaukee, USA)

- dithiothreitol (DTT) (Sigma, St. Louis, USA)

- Tris (TRIZMA base; Sigma, St. Louis, USA)

- guanidine hydrochloride (Sigma, St. Louis, USA)

- trifluoroacetic acid (TFA): HPLC grade

- acetonitrile: HPLC grade.
- buffer A (digestion buffer): 0.1 M Tris (pH 7.6)
- buffer B (alkylating buffer): 0.1 M Tris in 5 M guanidine-HCl and 1 mM EDTA, pH 7.6
- solvent A: 0.1% TFA in water
- solvent B: 0.08% TFA in acetonitrile

CNBr Cleavage

1. Dissolve protein sample in 100 µl 70% TFA.

2. Add 1 mg of CNBr or 2 µl of a 5 M solution of CNBr in acetonitrile. Purge with nitrogen/argon and leave to react for 18 h at room temperature in the dark.

3. Evaporate reaction mixture in SpeedVac.

4. Dissolve in 50 µl buffer B and inject into analytical column.

5. Separate by RP-HPLC using acetonitrile gradient: from 5%B to 40%B in 60 min at flow rate of 1 ml/min and temperature of 40 °C. Detect peptides at 210–214 nm, 0.1–0.2 aufs. Evaporate collected fractions in SpeedVac.

6. Dissolve each fraction in 40 µl buffer B. Add 1 µl (1.4 µmol/µl) of DTT and leave to react for 30 min at 37 °C.

7. Add 2 µl 4-vinylpyridine diluted 1:10 with ethanol (approx. 2 µmol) and leave to react for 10 min in the dark. Terminate reaction by injection into RP-column or add 1 ml of DTT as in step 6.

8. Separate alkylated peptides by RP-HPLC; use exactly the same gradient as in step 5 after washing out excess reagents and by-products.

9. Divide collected fractions into two equal portions and evaporate in SpeedVac.

10. Submit one part of alkylated fractions to amino acid and/or sequence analysis.

11. Find disulfide bonded fragment(s) by computer search or manual alignment.

12. Digest S-S-bonded fragments with trypsin as described below in case two or more disulfide bonds are present in a single fragment (S-S-core).

13. Dissolve S-S-core in 50 μl buffer A; add 1 μg trypsin and digest for 4 h at 37 °C.

14. Separate digestion mixture by RP-HPLC using acetonitrile gradient, e.g., from 0%B to 40%B at flow rate of 1 ml/min and temperature of 40 °C.

15. Collect fractions and follow steps 6–12.

17.5
Comments and Troubleshooting

- Filling the tip column with thiopropyl Sepharose 6B and manipulating it requires some experience. Practice before the experiment.

- Avoid reduction if protein does not contain disulfide bonds.

- Desalting of any protein samples by RP-HPLC is strongly recommended if the protein was purified by other methods.

- Store the reduced and desalted protein in a deep freezer (–70 °C).

- Thiopropyl Sepharose 6B may be regenerated two-three times after use.

- Since intact proteins containing S-S-bonds are poorly digested with enzymes, cleavage by chemical methods (e.g., CNBr cleavage) is recommended in order to isolate S-S-bonded fragments. Enzymes such as trypsin, chymotrypsin, Lys-C, Arg-C, and Glu-C may be used for subdigestion of disulfide-bonded fragments.

- Buffers not containing thiols (e.g., mercaptoethanol, DTT or cysteine), oxidizing agents, or heavy metals must be used for both protocols (except for when neccessary). Use high quality reagents and deionized water.

References

Egorov TA, Svenson A, Ryden L, Carlsson J (1975) Rapid and specific metod for isolation of thiol-containing peptides by thiol-disulfide exchange on a solid support. Proc Nat Acad Sci USA 72: 3029–3033

Egorov TA (1991) Use of Thiopropyl-Sepharose 6B for isolation and structure-functional analysis of thiol proteins. In: Jornvall H, Hoog J-O, Gustavsson A-M (eds) Methods in protein sequence analysis, Birkhausen, Basel, pp 177–185

Egorov TA, Musolyamov AKh, Andersen J, Roepstorff P (1994) The complete amino acid sequence and disulfide bond arrangement of oat alcohol-soluble avenin-3. Eur J Biochem 224: 631–638

Gray WR (1993) Disulfide structures of highly bridged peptides: a new strategy for analysis. Protein Science 2: 1732–1748

Chapter 18

Lipid Modifications of Proteins

T. G. GIANNAKOUROS

18.1
Introduction

Over the last 12 years or so, a group of posttranslational modifications of proteins with various lipid molecules has been identified in eukaryotic cells (for reviews see Schmidt 1989; McIlhinney 1990; Gordon et al. 1991; Magee 1991). In many cases these modifications have been shown to be essential for the subcellular localization of the proteins carrying them. Many intracellular proteins are now known to carry fatty acid modifications which serve such a purpose. The 14 carbon saturated fatty acid myristate (C14:0) is amide-linked to the NH_2-terminal of a range of proteins, exemplified by the $pp60^{src}$ proto-oncogene product. Site-directed mutagenesis has been used to demonstrate that this acylation is required both for membrane localization and transforming activity. The myristoylation enzyme has been isolated and is highly specific for acyl chain length. Inhibitors which interfere with this pathway provide an attractive approach to specific chemotherapies of cancers and viral infections including HIV.

Longer chain fatty acids, predominantly palmitate (C16:0), can be linked to proteins, usually via thioester bonds to cysteine or oxyester bonds to serine or threonine. Palmitoylation can occur both on transmembrane proteins and on entirely intracellular proteins. In the former case, however, the acylation sites are generally on the cytoplasmically disposed sequences or within the transmembrane region itself. Palmitoylation of ras proteins has been shown to be required for tight membrane binding and to cooperate with polyisoprenylation (see below) to determine specific subcellular localization.

Polyisoprenylation of proteins by alkylation of cysteine with 15 carbon (farnesyl) or 20 carbon (geranylgeranyl) steroid precursors has been identified over the last few years (for recent reviews see Magee and Newman 1992; Schafer and Rine 1992;

Sinensky and Lutz 1992). Farnesylation of ras proteins cooperates with nearby upstream palmitate residues or a polybasic sequence to specify plasma membrane binding. These modifications occur at COOH-terminal motifs of the general type CAAX (C, cysteine; A, aliphatic; X, any other amino acid). Small and hydrophilic side chains in the X position favor farnesylation of the cysteine residue, while large aliphatic residues favour geranylgeranylation. In addition to polyisoprenylation these sequences are also modified by proteolytic removal of the AAX sequence, and by carboxyl methylation of the α-carboxyl group. Geranylgeranyl addition can also be signaled by two other motifs, a CysCys or CysXCys sequence at the end of the protein, found in the ras- related family of rab proteins. The subtle variation in these lipid modifications, in concert with protein sequences, very likely contributes to the exquisite selectivity of different members of this family for compartments of the endomembrane system.

18.2.
Methods

18.2.1
Detection of Protein Acylation

In a few cases acylation has been recognized by structural analyses of unlabeled purified proteins. However, the modification of proteins with fatty acids is usually detected after metabolic labeling of appropriate cultured cell lines or primary cells with specific fatty acids (Magee and Courtneidge 1985; Magee et al. 1987). Due to the propensity of cells to interconvert fatty acids by chain elongation or shortening by β-oxidation, the identity of the incorporated label must be confirmed. Distinction between amide-linked and thioester-linked fatty acids can be made by simple chemical cleavages.

Metabolic Labeling of Tissue Culture Cells with Fatty Acids

Procedure 1. Grow cells in tissue culture dishes to 70%–90% confluence. Alternatively tissues can be rapidly removed from a newly killed animal, rinsed in serum-free tissue culture medium, and diced into small (approx. 1-mm^2) pieces with fine scissors and a scalpel. The pieces are transferred into appropriate dishes in complete tissue culture medium.

2. Replace the medium with warm fatty acid (FA) medium for 1 h. The composition of the FA medium is:

- Dulbecco's modified Eagle's medium (DMEM) (or suitable alternative depending on cell type
- 5% serum (newborn or fetal calf)
- 4x the normal concentration of nonessential amino acids
- 5 mM sodium pyruvate

3. Add [9,10-^{3}H] fatty acids in ethanol directly to the medium at a concentration of 50–200 µCi/ml (tilt the dish slightly and add the label to the "deep end" to prevent the ethanol from drying out the cells). Agitate gently to disperse the label.

4. Leave for the desired time (6 h–overnight).

5. Wash the cells with cold phosphate buffered saline (PBS) and lyse them directly in gel-loading buffer containing dithiothreitol or in immunoprecipitation buffer (RIPA): 20 mM Tris, 150 mM NaCl, 1 mM EDTA, 0.5% (v/v) NP40, 0.5% (w/v) Na-deoxycholate, 0.1% (w/v) SDS, 1% aprotinin (Sigma, St. Louis, USA) and 0.2 mM phenylmethyl-sulfonylfluoride (PMSF), pH 7.4.

6. Load the lysates directly or after immunoprecipitation with an appropriate antibody onto SDS-polyacrylamide gels. Run the dye front off the gel to remove free label and phospholipids. The gel is then dried and subjected to fluorography with 2,5-diphenyl-oxazole (PPO). PPO is more efficient than other water soluble fluors such as salicylate.

Comments and notes

- The use of a reduced serum concentration (1%–5%) and the inclusion of 5 mM Na pyruvate and increased amino acid concentrations in the labeling medium, were found to maximize incorporation of labeled fatty acids, whereas reincorporation of label was found to be significantly reduced.

- The 9,10-tritiated fatty acids provide the best combination of high specific activity and detectability for in vivo labeling. Also, due to the tritium being far removed from the carboxyl end where β-oxidation occurs, the reincorporation of label is minimized. For these reasons ^{14}C-fatty acids, especially those labeled in the 1 position, should be avoided for in vivo work. They can be used for in vitro studies of acylation.

- Fatty acids are often provided by the manufacturer as stocks in toluene. This should be removed under a gentle stream of nitrogen in a fume hood, and the fatty acids redissolved in ethanol at 10–50 mCi/ml. Under these conditions they can be stored for months or years at –20 °C.

- Since some thioester linked acyl groups are extremely labile and are removed even by reduction, the concentration of dithiothreitol in the gel loading buffer should not exceed 20 mM.

Linkage Analysis of Fatty Acid in Acylated Proteins

Distinction can be made between thioesters, oxyesters and amides by chemical cleavage with selective agents. This is most conveniently performed on duplicate slices from an SDS-polyacrylamide gel of [^{3}H]fatty acid-labeled protein(s). The method can be adapted for proteins in solution. Linkage analysis should not be performed on fixed or dried gels since both these procedures can lead to transacylation converting labile ester linkages to those of an unknown nature (possibly amides).

1. Excise replicate lanes from a fresh gel and shake at room temperature for 1 h with 0.2 M KOH in methanol (this cleaves thio- and oxyesters but not amides). Methanol alone is used to treat a control strip.

2. Treat a third strip with freshly prepared 1 M hydroxylamine hydrochloride titrated to pH 7.5 with NaOH (this rapidly cleaves thioesters but cleaves oxyesters or amides only poorly). Treat a fourth strip with 1 M Tris-HCl (pH 7.5) as a control. Hydroxylamine will cleave oxyesters at high pH (> 9).

3. Wash the slices three times with water for 5 min, and prepare them for fluorography as above.
   Cleavage is measured as a reduction in the fluorographic signal compared to controls and can be quantitated by densitometric scanning or scintillation counting of excised bands.

   Note: Hydroxylamine treatment in solution can also be used as a mild method for removal of thioester-linked fatty acid for functional studies or to prepare deacylated acceptor protein for in vitro acylation studies.

Analysis of protein bound label

1. Following electrophoresis, locate the band of interest either by using molecular weight standards, or by fluorography using sodium salicylate. PPO-treated gels cannot be used.

2. Excise the band from the wet or dried gel, and wash 3 × 5 min with 0.5 ml water (agitate), during which time the dried gel piece rehydrates and the salicylate is washed out.

3. Transfer the gel piece to a 1.5 ml microcentrifuge tube and lyophilize. Alternatively, protein can be digested out of the gel

with 200 mg/ml pronase in 20 mM ammonium bicarbonate, 0.05% SDS, pH 8.0 at 37 °C for 48 h, followed by lyophilization.

4. Place the microcentrifuge tube into a thick walled Teflon container with a tight fitting screw top. Add 1 ml 6 M HCl and flush tube with nitrogen. Close the lid tightly and hydrolyze in an oven at 110 °C for 16 h.

5. Extract the contents of the tube twice with 0.5 ml hexane and pool the extracts. Determine the radioactivity in the hexane extracts and in the residue (after dissolving in 1% SDS). Fatty acids should be quantitatively extracted into hexane, while label incorporated into sugars and amino acids will be mainly in the hexane residue.

6. Evaporate the hexane extracts (in a SpeedVac) and dissolve in a small volume (2–5 ml) of chloroform-methanol (2:1). Spot onto a RP18 thin-layer plate (Merck 13724, Merck, Darmstadt, Germany) and develop in acetonitrile-acetic acid (90:10). Spot authentic [^{3}H]fatty acids in parallel lanes as markers.

7. Dry the plate and detect the radioactivity by spraying with En^{3}hance spray (NEN, Hertfordshire, UK) and exposure to preflashed Kodak XAR-5 film at –70 °C, or by scraping 1 cm lengths of absorbent and scintillation counting.

18.2.2
Detection of Protein Prenylation

Polyisoprenoids such as farnesyl (15 carbon) and geranylgeranyl (20 carbon) are derived from mevalonic acid via a pathway which results ultimately in the production of a wide range of cellular metabolites, e.g., steroids, dolichols, ubiquinone, and heme (Goldstein and Brown 1990). The intermediates are produced as activated pyrophosphate derivatives which are polymerized starting from the 5 carbon precursor isopentyl pyrophosphate. The first step in the pathway is the production of mevalonic acid from hydroxylmethylglutaryl-coenzyme A (HMG-CoA) catalyzed by the enzyme HMG-CoA reductase. Several highly specific inhibitors of this enzyme are available (e.g., Mevinolin (lovastatin); Compactin (ML-236B) and Pravastatin) and are used clinically as anti-hypercholesterolemic agents.

Inhibition of HMG-CoA reductase by these agents causes depletion of cytosolic mevalonic acid, resulting in the accumulation of nonisoprenylated precursor proteins. The lack of isoprenylation frequently causes a decrease in hydrophobicity which can be assessed by phase partitioning using the detergent Triton X-114. [^{3}H]mevalonic acid added to treated cells is incorporated into the accumulated precursors, which can be detected as radiolabeled bands on SDS-PAGE. Having observed incorporation of label from [^{3}H]mevalonic acid into a protein it is important to determine the chain length of the polyisoprenoid moiety. Polyisoprenoids can be conveniently analyzed by HPLC after cleavage from the protein by methyl iodide (Casey et al. 1989).

Mevalonic Acid Labeling

R-[5-^{3}H]mevalonic acid is supplied in 50% aqueous ethanol. Just before use the ethanol should be removed by a gentle stream of nitrogen. The mevalonic acid must not be heated above 20 °C since it is unstable. Having removed the ethanol, the aqueous solution is added directly to cells, preincubated in FA medium containing 25–50 mM mevinolin for approximately 1 h, at a concentration of 50–200 µCi/ml for up to 16 h. At the end of the incubation the cells are processed for analysis by immunoprecipitation and SDS-PAGE (see also "Metabolic Labeling of Tissue Culture Cells with Fatty Acids").

For highest sensitivity PPO is used as a fluor. However if further analysis of the resolved radioactive bands is to be performed (see following section) fluorography is done with sodium salicylate. Typical exposure times are 1 week to 1 month.

HPLC Analysis of Polyisoprenoids

1. Excise salicylate-fluorographed bands from a dried gel, using the fluorogram as a guide. In order to have enough counts for analysis, a strong fluorographic signal should be obtained in 1 week. If longer exposures are required then multiple gel pieces should be used.

2. Place the gel piece in a 1.5 ml microcentrifuge tube and wash 3×5 min with double distilled water at room temperature to remove salicylate and SDS. Remove the paper backing from the gel at the same time.

3. Add 100 ml 0.1 M ammonium bicarbonate containing 0.02% sodium azide and 10 mg pronase to the gel piece. Incubate overnight at 37 °C. This digests the protein in the gel and aids elution.

4. Remove the supernatant and redigest with a further 100 ml ammonium bicarbonate and 10 mg pronase for 6 h at 37 °C.

Remove this supernatant and reextract for 2 h at 37 °C with 200 ml buffer alone.

5. Pool the three extracts in a 10 ml polypropylene tube and lyophilize. Dissolve the residue in 0.5 ml water and lyophilize again.

6. Dissolve the residue in 400 ml 25 mM Tris-HCl (pH 7.7) in 80% HPLC grade acetonitrile (degassed by nitrogen flushing to remove oxygen), and leave 1 h at room temperature. Add 800 ml N_2-degassed 3% (w/v) formic acid followed by 100 ml fresh methyl iodide. Flush with nitrogen and keep in the dark overnight at room temperature with vortex mixing.

7. Remove the methyl iodide under gentle vacuum (e.g., with a water pump), then add 150 ml 35% (w/v) sodium bicarbonate, flush with nitrogen and leave overnight in the dark with gentle mixing.

8. Extract the cleaved polyisoprenoids twice with 1.2 ml chloroform-methanol (9:1). Pool the extracts and count the organic extract and the aqueous residue.

9. Following cleavage, dry the chloroform-methanol pool under a stream of nitrogen and redissolve in 100 ml of HPLC grade acetonitrile containing 25 mM phosphoric acid (solvent B). It is important to use a high acetonitrile concentration since geranylgeraniol is poorly soluble in aqueous solvents. Just before HPLC analysis dilute the sample with an equal volume of HPLC grade water, vortex vigorously and microcentrifuge to remove particulate material.

10. Inject the sample (usually half) onto a reversed phase column (any octadecyl or shorter carbon chain can be used, with appropriate optimization of the separation) equilibrated in 50% aqueous acetonitrile containining 25 mM phosphoric acid (solvent A) at a flow rate of 0.5 ml/min. Solvent B is 100% acetonitrile/25 mM phosphoric acid. The gradient used for elution is: 10 min with 100% solvent A, linear gradient to 100% solvent B over 10 min, 10 min hold at 100% solvent B, 5 min reverse gradient to 100% solvent A.

Collect 1 min fractions in scintillation vial inserts and count after the addition of 5 ml scintillation fluid. Standard isoprenoids used as markers are the following: geraniol (Sigma, St. Louis, USA), farnesol (Sigma, St. Louis, USA or Aldrich, St. Louis, USA), geranylgeraniol (Kuraray Co. Ltd., Kyoto, Japan).

References

Casey PJ, Solski PA, Der CJ, Buss JE (1989) p21ras is modified by a farnesyl isoprenoid. Proc Natl Acad Sci USA 86: 8323–8327

Goldstein JL, Brown MS (1990) Regulation of the mevalonate pathway. Nature 343: 425–430

Gordon JI, Duronio RJ, Rudnick DA, Adams SP, Gokel GW (1991) Protein N-myristoylation. J Biol Chem 266: 8647–8650

Magee AI (1990) Lipid modification of proteins and its relevance to protein targeting.J Cell Sci 97: 581–584

Magee AI, Courtneidge SA (1985) Two classes of fatty acylated proteins exist in eukaryotic cells. EMBO J 4: 1137–1144.

Magee AI, Gutierrez L, McKay IA, Marshall, CJ, Hall A (1987) Dynamic fatty acylation of p21^{N-ras}. EMBO J 6: 3353–3357

Magee T and Newman C (1992) The role of lipid anchors for small G proteins in membrane trafficking. Trends Cell Biol 2: 318–323

McIlhinney RAJ (1990) The fats of life: the importance and function of protein acylation.Trends Biochem Sci 15: 387–391

Schafer WR, Rine J (1992) Protein prenylation: genes, enzymes, targets and functions. Annu Rev Genet 30: 209–237

Schmidt MFG (1989) Fatty acylation of proteins. Biochim Biophys Acta 988: 411–426.

Sinensky M, Lutz RJ (1992) The prenylation of proteins. BioEssays 14: 25–31

Part VII
Mass Spectrometry in Peptide and Protein Analysis

Mass Spectrometry in Peptide and Protein Sequence Analysis

B. THIEDE

19.1
Introduction

For many years mass spectrometry has been known as the most accurate and sensitive method of obtaining molecular weight information for small molecules. Revolutionary developments in ionization techniques and instrumentation have recently demonstrated that these capabilities can be extended to molecules much larger than 10 kDa. For this purpose, the "soft" ionization processes, matrix-assisted laser desorption/ionization (MALDI) and electrospray ionization (ESI), appear particularly promising. The accuracy, sensitivity and resolving power of these techniques permit the detection of proteins and other biopolymers. Structural information of peptides and proteins by mass spectrometry can be obtained by tandem mass spectrometry (MS/MS) and post-source decay matrix-assisted laser desorption/ionization (PSD-MALDI-MS).

Electrospray Ionization Mass Spectrometry (ESI-MS)

The ESI phenomenon is a process that produces naked, intact molecules in ionized form from an analyte solution. A spray of fine, highly charged droplets is created at atmospheric pressure in the presence of a strong electric field. The electrospray source at elevated voltage relative to a counter electrode with an orifice where ions entrained in a flow of gas enter the mass spectrometer. Liquid flow is generated by infusion syringes, separation devices (high performance liquid chromatography, HPLC; capillary electrophoresis, CE) or other liquid sources at flow rates usually between one and a few microliters per minute. The resulting field between capillary tip and the interface plate charges the surface of the emerging liquid, dispersing it by Coulomb forces into a fine spray of charged droplets. The so formed fine droplets carry an excess of charge and are attracted to the inlet of the mass spectrometer, which is held at a lower potential. A counter-current flow of dry gas to the droplets causes evaporation of solvent from each droplet, decreasing its diameter. Consequently, the charge density on its surface increases until the so-called Rayleigh limit is reached, at which the Coulomb repulsion becomes of the same order as the surface tension. The resulting instability, sometimes called a Coulomb explosion, tears the droplet apart, producing charged daughter droplets that also evaporate. This sequence of events repeats and finally produces droplets so small that the combination of charge density and radius of curvature at the droplet surface produces an electric field intense enough to finally desorb ions from the droplets into the gas phase. At last this ion desorption mechanism produces "quasi-molecular" ions suitable for mass analysis. The resulting naked molecules, each carrying an excess of positive charges related to its charge state in the original solution, are directed into a quadrupole filter for m/z ratio analysis (Fenn 1989).

The development of the nanoelectrospray (nanoES) ion source coupled with a rapid desalting/concentration step allows the analysis of impure samples such as peptide mixtures extracted after in-gel digestion (Wilm 1996).

Mass accuraccies of about 0.01% and resolutions of about 1000 are routinely obtained in ESI-MS down to a level of 10 fmol in 5–10 min. Electrospray has proven successful for the molecular weight determination of 200 kDa molecular ions with quadrupole instruments limited to m/z 2500. It is optimal for on-line coupling with separation techniques such as liquid chromatography and capillary zone electrophoresis.

Matrix-Assisted Laser Desorption/Ionization Mass Spectrometry (MALDI-MS)

For MALDI-MS the samples are prepared by adding the analyte molecule to a concentrated aqueous solution containing a large molar excess of a matrix material such as dihydoxybenzoic acid and a variety of cinnamic acid derivatives, followed by drying of the solution on the probe tip and insertion into the mass spectrometer (Hillenkamp 1991). Another preparation method is the thin-layer technique in which the matrix evaporates fast as a thin layer and then the analyte solution is deposited on the layer (Vorm 1994).

In the mass spectrometer the sample is bombarded with short duration (1–10 ns) pulses at UV wavelength. The suitable matrices exhibit a strong absorption at this wavelength. The interaction with the laser pulse causes desorption and ionization of the matrix and the analyte molecule. A strong electric field is imposed upon ions generated from the sample by application of a high potential (typically 25–30 kV) to the sample probe with respect to a closly spaced accelerating electrode. Analyte ions are thus accelerated through the orifice in the electrode and enter in a field-free flight-tube (typical length: 50–200 cm). The masses of the analyte ions can be simply determined by time-of-flight (TOF) analysis. As the ion passes through the field-free flight-tube, they separate into series of spatially discrete individual ion packets, each traveling with a velocity proportional to $m/z^{1/2}$. A detector positioned at the end of the flight-tube produces a signal as each ion packet strikes it. The difference between the start time, set by the occurrence of the laser pulse common to all ions, and the arrival time of an individual ion at the detector can be used to calculate the ion's m/z ratio.

Besides linear instruments, reflectron mass spectrometers are used in order to compensate initial energy spreads. In these instruments an ion mirror is placed in the flight path. The peak width contribution from the initial velocity distribution can be largely corrected and this leads to an increased mass resolution for stable ions.

Molecular masses much greater than 300 kDa have been successfully analyzed by MALDI-MS and the practical mass range seems to be unlimited. Only femtomoles of a sample are required in the analysis which takes less than 5 min to perform. Mass accuracy is dependent of the size of the analyte and varies between 0.01% and 0.1%. MALDI-MS is an excellent technique for the analysis of complex mixtures. Furthermore it is very tolerant of the presence of salt and other contaminants in the sample.

New developments such as MALDI-Fourier transform mass spectrometry (Castoro 1992) and delayed extraction (Brown 1995) are viable methods for obtaining high mass accuracy and high mass resolution.

Tandem Mass Spectrometry (MS/MS)

Another invention in mass spectrometry is the use of a tandem instrument, in which two mass spectrometers are hooked together in series to characterize biological molecules in mixtures without prior separation of the components (Biemann 1987; Hunt 1986). In the first mass spectrometer the protonated molecular ions are generated. In a mixture a precursor ion can then be selected and fragmentation can be achieved by collision-induced decomposition (CID), in which the ion collides with a neutral atom (i.e., helium), or by other methods, such as photon excitation. The product ions are then analyzed in the second mass spectrometer.

Post-Source Decay - Matrix-Assisted Laser Desorption/Ionization Mass Spectrometry (PSD-MALDI-MS)

The tendency of laser-desorbed mass ions to undergo post-source decay can be used to elucidate structures of peptides and proteins (Spengler 1991). Peptides and proteins are rather unstable on their way through the TOF instrument. This metastable decay of matrix-desorbed ions can be detected in reflectron instruments. Therefore the electric potential of the reflectron has to be adjusted according to the remaining kinetic energy of the metastable fragment ion. In order to obtain the complete mass range of these ions, a series of spectra need to be acquired by reducing the reflectron potential stepwise from its nominal value down to zero. Distinct ions in a mixture can be selected without separation by the means of an ion gate. Furthermore the additional introduction of a collision cell in a MALDI-MS allows the distinction between leucine and isoleucine.

Sequencing of Unknown Peptides and Proteins

The amino acid sequences of peptides and proteins can be conventionally determined directly by stepwise Edman degradation and indirectly by DNA sequencing. Limitations of the Edman degradation are the inability to sequence cyclic peptides, amino-terminally blocked peptides and proteins, and the detection of certain modified amino acids. Post-translational modifications cannot be evaluated by base sequencing of the gene.

These problems can be solved by applying MS/MS or PSD-MALDI-MS. For protein sequencing the protein must be digested with specific proteases or chemicals to generate peptides. The sequence will be completely interpretable in many resulting spectra. Problems can occur by the absence of one or more complete ion series, the lack of fragmentation between the two amino-terminal and/or two carboxy-terminal amino acids, the absence of "secondary" ions to distinguish leucine and isoleucine, the formation of peptides longer than 20–30 amino acids, or the appearance of peptides with the same mass as the matrix ions. Furthermore the alignment of the individual peptides in the protein chain is unknown. For all these reasons, one or more additional digests are needed to sequence a protein.

Two methods for protein sequencing, the so-called ladder sequencing, have been developed by combining Edman degradation and MALDI-MS (Chait 1993; Bartlet-Jones 1994). The method of Chait used Edman degradation with 5% phenylisocyanate, a terminating reagent, to form the phenylcarbamyl derivative of the whole peptide and 95% of the normally used phenylisothiocyanate to form the shortened peptide after cleavage with trifluoroacetic acid. Successive Edman degradation cycles were performed with the remaining degraded peptides. Finally the mixture of the phenylcarbamyl peptides was analyzed in one step by MALDI-MS. The approach of Bartlet-Jones used volatile reagents, trifluoroethylisothiocyanate for the coupling and heptafluorobutyric acid for the cleavage, to avoid peptide loss through extraction. A nested set of peptides were generated by adding equal aliquots of fresh peptide to each cycle and the coupling and cleavage reactions were driven towards completion. Finally, one MALDI-MS spectra revealed the sequence. The methods are limited by a mass accuracy of 0.01% for peptides up to 4000 Da to avoid ambiguities in the identification of amino acids with a mass difference of 1 Da. Furthermore leucine and isoleucine cannot be distinguished by mass measuring alone.

Another strategy for sequencing unknown proteins combining mass spectrometry, especially MALDI-MS, with Edman degradation is based on a time course monitoring of an enzymatic or chemical cleavage by MS with starting amounts of 1 picomol protein (Roepstorff 1993). A preparative digest is then produced and the peptides are separated by HPLC and analyzed by MS and/or by amino acid analysis, followed by Edman degradation. The molecular weight determination of the peptides yields an estimation of the number of cycles to perform and confirms the sequencing result; modified amino acid residues can also be determined. Furthermore, by adding the molecular weights of the peptides and comparing them with the molecular weight of the intact protein, peptides which correspond to the complete sequence can be selected for sequencing. At this stage most of the sequence will have been established. An alternative digestion can be carried out by employing the same strategy. The nature of amino-terminal blocking groups can be identified by comparison of amino acid analysis with the mass determination.

Carboxy-terminal sequence information can be obtained by combination of carboxypeptidase-digestion with MALDI-MS by identifying the truncated peptides (Thiede 1995).

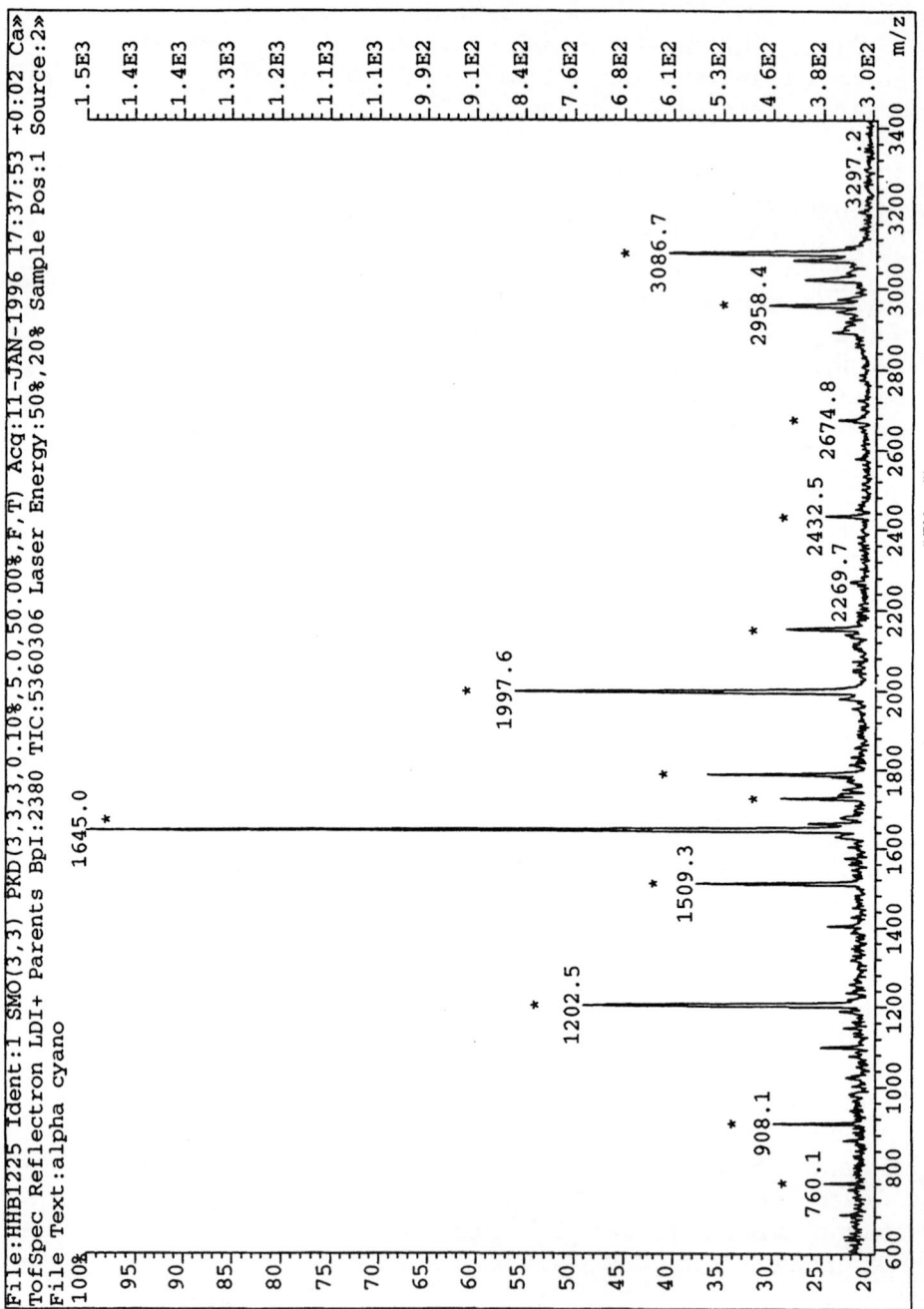

Fig. 1. MALDI-MS spectrum of the Lys-C digest of creatine kinase M-chain. The protein was derived from a 2-DE gel of the human myocardium (Jungblut 1994). The 13 peaks marked with *asterisks* matched with peptides of the creatine kinase M-chain and identified the protein unequivocally

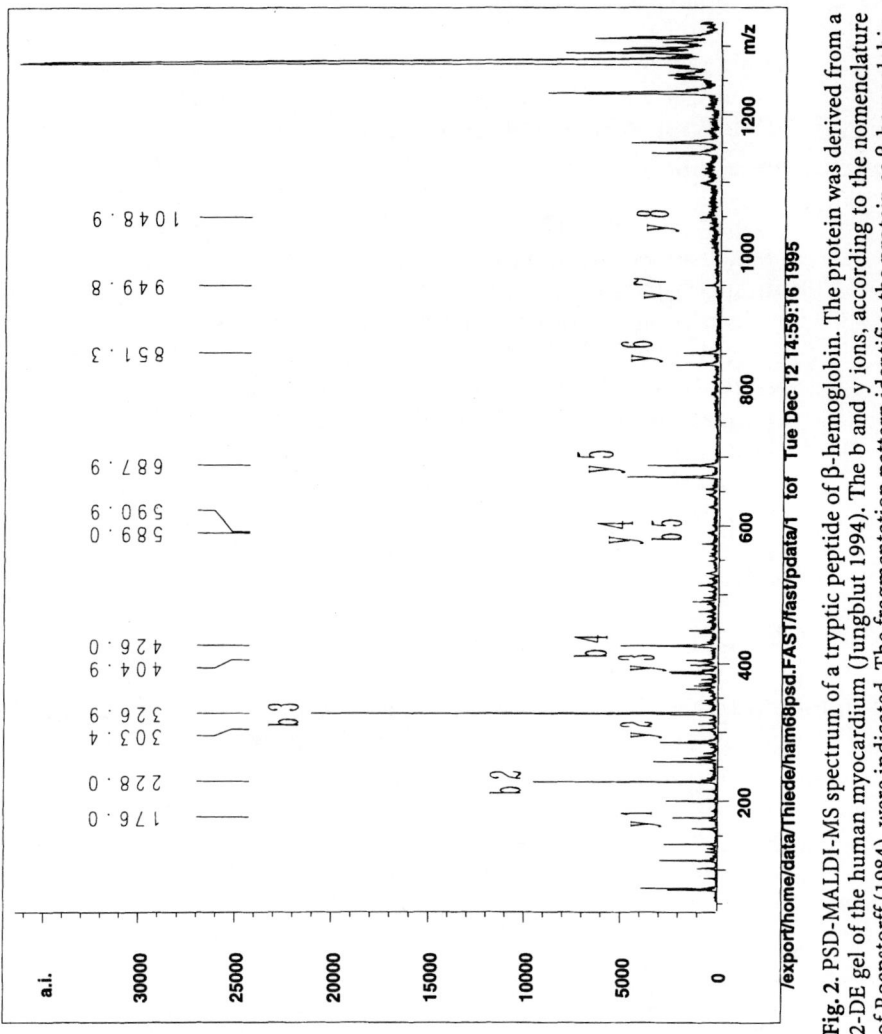

Fig. 2. PSD-MALDI-MS spectrum of a tryptic peptide of β-hemoglobin. The protein was derived from a 2-DE gel of the human myocardium (Jungblut 1994). The b and y ions, according to the nomenclature of Roepstorff (1984), were indicated. The fragmentation pattern identifies the protein as β-hemoglobin

Identification of Known Proteins

The question whether a protein is known or related to a known protein is a fundamental question in protein research. Protein sequences, derived either by DNA sequencing or protein sequencing, are recorded in protein sequence databases. The total molecular weight can easily be determined by ESI-MS and MALDI-MS and used to search in a database (Mann 1993). As a result a presumption can be made which subsequently must be pursued further.

Identification of proteins by molecular mass mapping of peptide fragments derived from enzymatic or chemical cleavage in sequence databases yields more specificity. Furthermore this application can be performed on proteins derived from 2D-gel electrophoresis (Henzel 1993) (Fig. 1). Additional sequence information of selected peptides in the mixture can be generated by PSD-MALDI-MS (Fig. 2) and nanoES-MS/MS (Fig. 3). Furthermore the sequence-tag method can be used for identification with only two or three amino acids (Mann 1994).

Selected Post-translational Modifications

Phosphorylation   Phosphorylation of the hydroxyl groups of specific serine, threonine, and/or tyrosine residues is a common post-translational modification of proteins. In eukaryotes, protein phosphorylation-dephosphorylation is one of the most important mechanisms for regulating many intracellular functions including signal transduction and cell division.

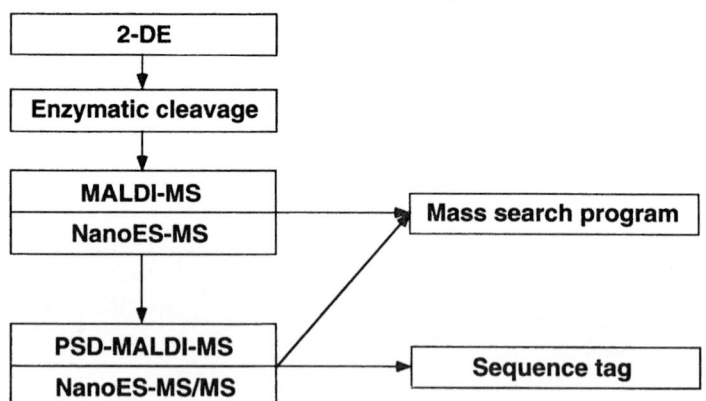

Fig. 3. Identification of proteins from 2-DE gels by mass spectrometry without separation by HPLC

Several methods have been developed and used to locate phosphorylation sites in proteins. All these methods apply the strategy of degrading the phosphoprotein chemically or enzymatically into small peptides, followed by analysis of the composition and sequence of each fragment. Phosphopeptide mapping has been used by sodium dodecyl sulfate-polyacrylamide gel electrophoresis (SDS-PAGE), reverse-phase HPLC, two-dimensional separation on TLC-plates or automated Edman sequence analysis.

MALDI-MS and ESI-MS have also been used to identify phosphorylation sites (Liao 1994). After degradation of the phosphoprotein into small peptides by specific enzymatic or chemical reactions, the identification of the phosphopeptides is obtained by −80 Da (or multiples of −80 Da) mass shifts in the mass spectra prior to and after dephosphorylation with alkaline phosphatase (Fig. 4). If the primary sequence of the protein is known the phosphorylation site can be located by mass mapping of peptide fragments. The digestion mixture can be directly analyzed by MALDI-MS or it can be fractionated by reverse-phase HPLC and then subjected to MS.

The advantages of using MALDI-MS or ESI-MS in order to identify and locate phosphorylation sites in phosphoproteins in comparison to other methods are the high mass accuracy (< 0.1%), the ability to analyze mixtures and the low time consumption. Furthermore the high sensitivity of these methods is important if the amount of available phosphoprotein is limited. In many cases the techniques described allow the use of multiple enzymatic digests to obtain the desired information. If no degradation is possible between two or more putative phosphorylation sites the sequence position can be determined by CID or PSD.

Glycosylation

Attachment of carbohydrates is one of the most prevalent post-translational modifications of eukaryotic proteins. There are two types of glycosylation, N-type and O-type, depending on the atom of the protein to which the carbohydrate is attached. N-type glycosylation occurs exclusively on the nitrogen atom of the asparagine side chain, whereas O-glycosylation occurs on the oxygen atoms of hydroxyls, particularly those of serine and threonine residues. A description of glycosylation is rendered more difficult by the complexity of the carbohydrate structures attached to the proteins. At least eight different sugar monomers are used and joined by a variety of glycoside linkages between their various functional groups.

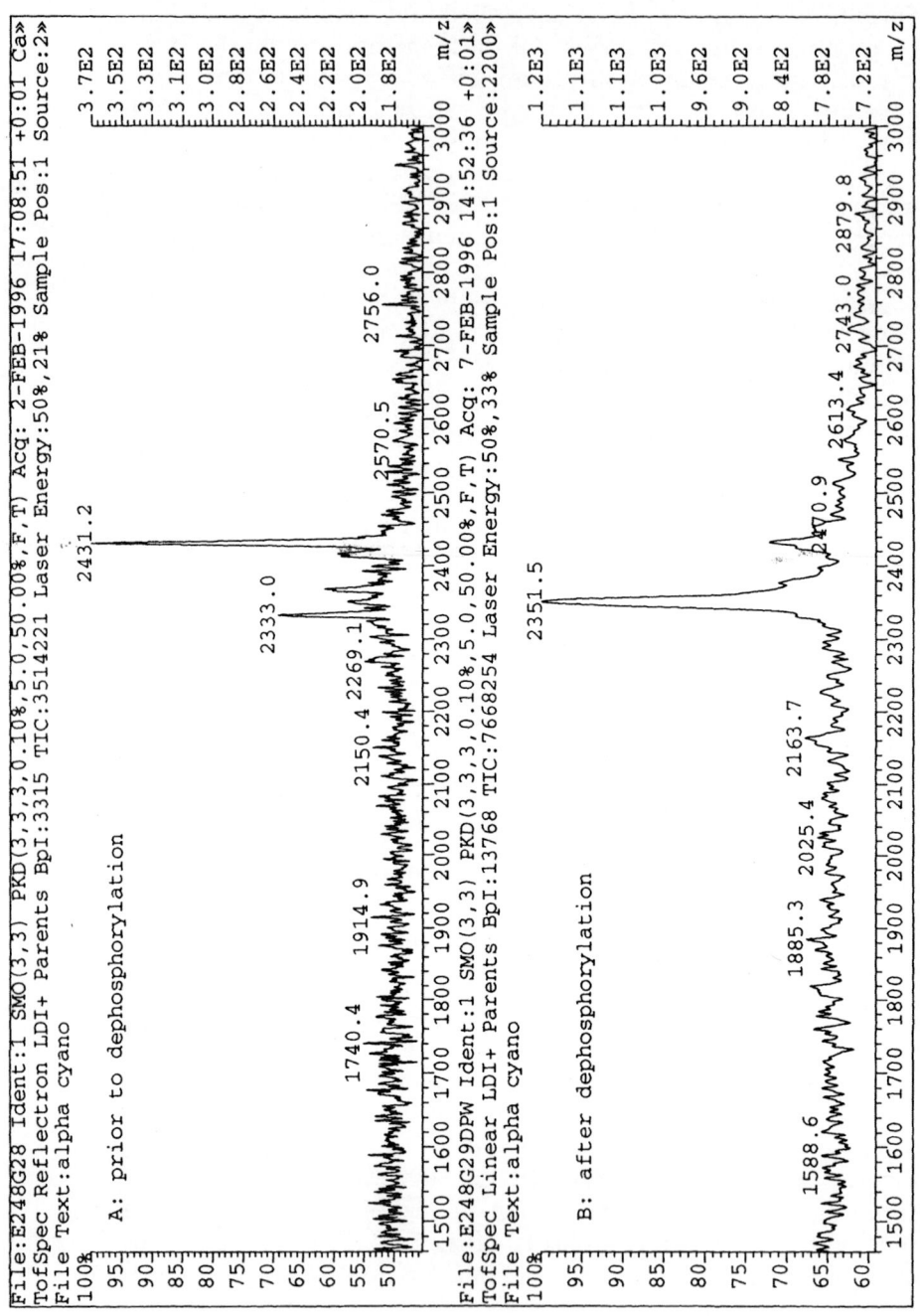

Fig. 4A,B. MALDI-MS spectra from a Lys-C peptide of myosin light chain 2, A prior to and B after dephosphorylation with alkaline

The molecular mass detection of glycopeptides is the prerequisite for the determination of the carbohydrate composition. For this purpose the glycoprotein is enzymatically digested. Chemical clevage must be used with extreme caution since the glycosidic linkages may be labile under the cleavage conditions. The carbohydrate portion of the glycopeptide is obtained by measuring the molecular masses before and after treatment with specific glycosidases. Coupling the obtained masses with database information faciliates determination of the carbohydrate composition since there is limited mass degeneracy for most compositions. Nevertheless the anomeric and linked isomers as well as the various forms of hexoses and hexosamines are indistinguishable by mass measurement alone.

ESI-MS and MALDI-MS have been used to characterize glycoproteins with amounts in the subpicomole level. MALDI-MS is an ideal technique for monitoring molecular mass shifts resulting from several specific glycosidase digestions because of the high insensitivity to buffers. Furthermore the comparison of linear and reflector mode spectra shows shifts in masses in many events (Huberty 1993). This observation reflects the metastable decay of silalyated glycopeptides or the loss of most neuraminic acid residues; loss of hexoses and hexosamine has also been observed. By stepwise reducing the reflector potential in a PSD experiment fragment ions associated with the structure of the peptide and oligosaccharide can be generated. Furthermore, MS/MS has been used to characterize glycopeptides (Richter 1990). By investigating the full glycopeptide or the separated peptide and saccharides or oligosaccharides after glycosidase digestion in combination with chemical derivatization more information can be evaluated, especially about the oligosaccharide content.

19.2
Materials

- matrix-assisted laser desorption/ionization mass spectrometer (VG TofSpec, Fisons Instruments, Manchester, UK, and Bruker-Reflex, Bruker-Franzen-Analytik, Bremen, Germany) **Equipment**

- vortex (Bender and Hobein, Zurich, Switzerland)

- centrifuge (Centrifuge 5415C, Eppendorf-Netheler-Hinz GmbH, Hamburg, Germany)

- matrix 1: 50 mM α-cyano-4-hydroxycinnamic acid (CCA; Sigma, Deisenhofen, Germany), 40% acetonitrile (Merck, Darmstadt, Germany), 60% aqueous 0.1% trifluoroacetic acid **Reagents**

(Merck, Darmstadt, Germany). Vortex and centrifuge for 2 min at 12000 rpm.

Note: CCA is a saturated solution; make fresh weekly and store at 4 °C.

- matrix 2: 50 mM α-cyano-4-hydroxycinnamic acid (CCA), 99% acetone (Merck, Darmstadt, Germany), 1% aqueous 0.1% trifluoroacetic acid. Vortex and centrifuge.

Note: CCA is a saturated solution; make fresh weekly and store at 4 °C.

- matrix 3: 50 mM sinapinic acid (SA; Fluka, Neu-Ulm, Germany), 30% acetonitrile, 70% aqueous 0.1% trifluoroacetic acid. Vortex and centrifuge.

Note: SA is a saturated solution. Make fresh every day and store at 4 °C, M+206 sinapinic acid adduct signals.

- matrix 4: 100 mM 2,5 dihydroxybenzoic acid (DHB; Sigma, Deisenhofen, Germany), 30% acetonitrile, 70% aqueous 0.1% trifluoroacetic acid. Vortex and centrifuge.

Note: Make fresh DHB weekly and store at 4 °C.

19.3
Preparation Methods for MALDI-MS

Dried drop method

1. Dissolve the sample.

   Note: Most peptides and proteins are soluble in 30% acetonitrile/70% aqueous 0.1% trifluoroacetic acid.

2. Vortex.

3. Short centrifugation.

4. Place matrix 1 or 3 or 4 on the target.

   Note: CCA is mainly used for peptides up to 10 kDa. SA gives good results with peptides and proteins larger than 3 kDa while DHB is recommanded for glycoproteins.

5. Add sample to the matrix on the target.

   Note: Ratio of amount of matrix solution to amount of sample-solution should be about 3:2–1:1.

6. Introduce the dry target into the mass spectrometer.

7. Record the mass spectra.

1. Dissolve the sample.

   Note: Most peptides and proteins are soluble in 30% acetonitrile/70% aqueous 0.1% trifluoroacetic acid.

2. Vortex.

3. Short centrifugation.

4. Place matrix 2 on the target.

   Note: The amount of matrix 2 must be optimized for each different target; it should dry in 10 s maximum.

5. Add the sample onto the dry matrix.

6. After drying wash the sample three times by placing 1 μl water on the top and blow off the water (e.g., by pressurized air).

7. Introduce the dry target into the mass spectrometer.

8. Record the mass spectra.

Thin-layer method

19.4
Results and Comments

The identification of proteins, e.g., from 2-DE gels can be performed by peptide mass fingerprinting (Fig. 1); nevertheless, identification is not unequivocal in any case. For such a purpose a second identification method is necessary. Sequence information can be obtained by performing PSD-MALDI-MS with one peptide of the peptide mass fingerprint by using the same sample (Fig. 2). Mass spectrometry is the method of choice for the identification of post-translational modifications (Fig. 3).

19.5
Troubleshooting

- Avoid the use of nonvolatile reagents such as salts (e.g., NaCl, KH_2PO_4, $CaCl_2$), detergents (e.g., Triton, SDS), solvents (e.g., glycerol, DMSO). If these reagents cannot be avoided purify the sample by, e.g., RP-HPLC. Sometimes dilution of the sample results in obtainable spectra. The volatile reagents, e.g., ammonium hydrogen carbonate, ammonium acetate, acetonitrile, trifluoroacetic acid, are suitable for MALDI-MS. As detergent, 1% β-octylglycopyranoside can be recommended.

- Only femtomoles of a sample are required for MALDI analysis. Nethertheless the amount of sample needed depends on the sample type (peptide, protein, glycoprotein), molecular weight (the larger the more needed) and (very important) purity.

- Work whenever possible in the reflectron mode. For the investigation of mixtures, smaller peaks are sometimes more readily detectable in the linear mode.

- Matrix ions can be used to check the calibration (CCA: 190.1/379.1; SA: 225.1/449.1; DHB: 155.0/309.1).

References

Bartlet-Jones M, Jeffery WA, Hansen HF, Pappin DJC (1994) Peptide ladder sequencing by mass spectrometry using a novel, volatile degradation reagent. Rapid Commun Mass Spectrom 8:737–742

Biemann K, Scoble HA (1987) Characterization by tandem mass spectrometry of structural modifications in proteins. Science 237:992–998

Brown RS, Lennon JJ (1995) Sequence-specific fragmentation of matrix-assisted laser-desorbed protein/peptide ions. Anal Chem 67:3990–3999

Castoro JA, Köster C, Wilkins CL (1992) Matrix-assisted laser desorption/ionization high-mass molecules by Fourier-transform mass spectrometry. Rapid Commun Mass Spectrom 6:2621–2627

Chait BT, Wang R, Beavis RC, Kent SBH (1993) Protein ladder sequencing. Science 262:89–92

Fenn JB, Mann M, Meng CK, Wong SF, Whitehouse CM (1989) Electrospray ionization for mass spectrometry of large biomolecules. Science 246:64–71

Henzel WJ, Billeci TM, Stults JT, Wong SC, Grimley C, Watanabe C (1993) Identifying proteins from two-dimensional gels by molecular mass searching of peptide fragments in protein sequence databases. Proc Natl Acad Sci 90:5011–5015

Hillenkamp F, Karas M, Beavis RC, Chait BT (1991) Matrix-assisted laser desorption/ionization mass spectrometry of biopolymers. Anal Chem 63:1193A-1203A

Huberty MC, Vath JE, Yu W, Martin SA (1993) Site-specific carbohydrate identification in recombinant proteins using MALD-TOF MS. Anal Chem 65:2791–2800

Hunt DF, Yates III JR, Shabanowitz J, Winston S, Hauer CR (1986) Protein sequencing by tandem mass spectrometry. Proc Natl Acad Sci 83:6233–6237

Jungblut P, Otto A, Zeindl-Eberhart E, Pleißner KP, Regitz-Zagrosek V, Fleck E, Wittmann-Liebold B (1994) Protein composition of the human heart: The construction of a myocardial two-dimensional electrophoresis database. Electrophoresis 15:685–707

Liao PC, Leykam J, Andrews, PC, Gage DA, Allison J (1994) An approach to locate phosphorylation sites in a phosphoprotein: mass mapping by combining specific enzymatic degradation with matrix-assisted laser desorption/ionization mass spectrometry. Anal Biochem 219:9–20

Mann M, H¢jrup P, Roepstorff P (1993) Use of mass spectrometric molecular weight information to identify proteins in sequence databases. Biol Mass Spectrom 22:338–345

Mann M, Wilm, M (1994) Error-tolerant identification of peptides in sequence databases by peptide sequence tags. Anal Chem 66:4390–4399

Richter WJ, Müller DR, Domon B (1990) Tandem mass spectrometry in structural characterization of oligosaccharide residues in glycoconjugates. Meth Enzym 193:607–623

Roepstorff P, Fohlmann J (1984) Proposal for a common nomenclature for sequence ions in mass spectra of peptides. Biomed Mass Spectrom 11:601

Roepstorff P, Højrup P (1993) A general strategy for the use of mass spectrometric molecular weight determination in protein purification and sequence determination. In: K. Imahori, F. Sakiyama (eds) Methods in protein sequence analysis. Plenum, New York pp 149–156

Spengler B, Kirsch D, Kaufmann R (1991) Metastable decay of peptides and proteins in matrix-assisted laser-desorption mass spectrometry. Rapid Commun Mass Spectrom 5:198–202

Thiede B, Wittmann-Liebold B, Bienert M, Krause E (1995) MALDI-MS for C-terminal sequence determination of peptides and proteins degraded by carboxypeptidase Y and P. FEBS Lett 357:65–69

Vorm O, Roepstorff P Mann M (1994) Improved resolution and very high sensitivity in MALDI TOF of matrix surfaces made by fast evaporation. Anal Chem 66:3281–3287

Wilm M, Shevchenko A, Houthaeve T, Breit S, Schweigerer L, Fotsis T, Mann M (1996) Femtomole sequencing of proteins from polyacrylamide gels by nano-electrospray mass spectrometry. Nature 379:466–469.

Chapter 20

Carboxy-Terminal Sequencing by Combining Carboxypeptidase Y and P Digestion with Matrix-Assisted Laser Desorption/Ionization Mass Spectrometry

B. THIEDE

20.1
Introduction

Carboxy-terminal sequence information is of particular interest for: (1) investigating amino-terminal blocked peptides and proteins, (2) identification of proteins derived from two-dimensional polyacrylamide gel electrophoresis (2-DE), (3) confirmation of DNA sequence data, (4) cloning experiments and (5) recombinant protein characterization. Chemical methods for carboxy-terminal sequencing have been developed (reviewed in Inglis 1991); however, no method with results comparable to those obtained with Edman degradation have yet been described for amino-terminal sequencing (Edman 1967). Carboxypeptidases have been used frequently for carboxy-terminal sequencing either by determination of the released free amino acids or by identification of the truncated peptides by mass spectrometry. Matrix-assisted laser desorption/ionization mass spectrometry (MALDI-MS) (Hillenkamp 1991) is an optimal technique for the investigation of peptide mixtures derived from enzymatic digests (Henzel 1993) because of the sensitive detection in the femto-mole-picomole range and the tolerance towards many buffer systems. Carboxypeptidases have different preferences for each amino acid and each individual peptide bond (reviewed in Breddam 1986). The combined use of carboxypeptidase Y and carboxypeptidase P and mass measurement by MALDI-MS of samples taken after different digestion times can generate carboxy-terminal truncated peptides of all amino acids (Fig. 1) (Thiede 1995).

Peptide mass fingerprinting is a powerful tool for the identification of proteins, especially for proteins derived from 2-DE (Henzel 1993; Pappin 1993). Nevertheless, an unequivocal identification is sometimes complicated. A more confident identification without the use of more material can be achieved by additional mass measurement with MALDI-MS of the micro-HPLC

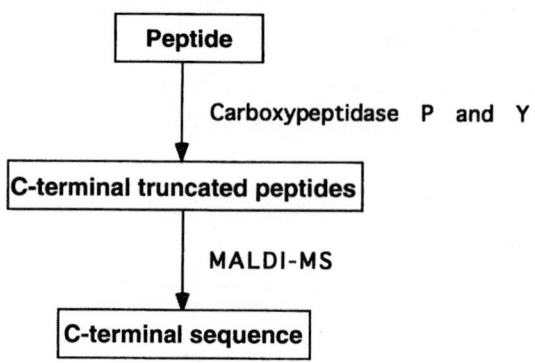

Fig. 1. Flow-chart demonstrating the carboxy-terminal sequencing of peptides by combining carboxypeptidase P and Y digestion and MALDI-MS

separated peptides prior to and after on-target treatment with carboxypeptidases P and Y. This approach can be used for the identification of proteins, especially if the amount of material is limited and less sensitive methods are not applicable.

20.1
Materials

Equipment – matrix-assisted laser desorption/ionization mass spectrometer, e.g., FISONS Instruments TofSpec (Manchester, UK)

Reagents – matrix 1: 50 mM α-cyano 4-hydroxycinnamic acid (Sigma, Deisenhofen, Germany), 50% acetonitrile (Merck, Darmstadt, Germany), 50% aqueous 0.1% trifluoroacetic acid (Merck, Darmstadt, Germany)

– matrix 2: 50 mM α-cyano-4-hydroxycinnamic acid, 99% acetone (Merck, Darmstadt, Germany), 1% aqueous 0.1% trifluoroacetic acid

– digestion buffer: 10 mM sodium citrate (pH 5.0), carboxypeptidase P and carboxypeptidase Y (both sequencing grade from Boehringer-Mannheim, Mannheim, Germany)

20.3
Experimental Procedure

1. Dissolve the sample in 8 µl digestion buffer.

   Note: A control peptide (e.g. 40 pmol substance P, Bachem, Heidelberg, Germany) should be used simultaneously to ensure the quality of the carboxypeptidases and to check the calibration of the mass spectrometer.

2. Add 1.2 µl matrix 1 into eight tubes.

3. Add 0.8 µl sample solution into one of the eight tubes.

4. Add 1.0 µl carboxypeptidases Y and P (1:1) to the sample solution. The ratio of enzyme to sample should be about 1:50 by weight.

5. Remove a 0.8 µl aliquot of the digestion mixture after 1, 2, 5, 10, 15, 30 and 60 min, placing each aliquot into a separate matrix 1-containing tube.

   Note: Matrix 1, with a pH < 2 serves as the stop solution.

6. Place each mixture (2 µl) onto the target.

   Note: The amount to be placed onto the target depends on the size of the target.

7. Introduce the target into the mass spectrometer after air drying.

8. Record the mass spectra of the eight samples.

9. Compare the spectra and calculate the mass differences from the highest (the parent peptide) to the lowest mass. The mass differences allow the determination of the cleaved amino acids from the carboxy-terminal (glycine 57.1, alanine 71.1, serine 87.1, proline 97.1, valine 99.1, threonine 101.1, cysteine 103.2, leucine 113.2, isoleucine 113.2, asparagine 114.2, aspartic acid 115.1, glutamine 128.1, lysine 128.2, glutamic acid 129.1, methionine 131.2, histidine 137.2, phenylalanine 147.2, arginine 156.2, tyrosine 163.2 and tryptophan 186.2).

   Note: The differentiation between leucine/isoleucine and glutamine/lysine is not possible by MALDI-MS. The distinction between some amino acids depends on the mass accuracy.

Time monitoring of carboxypeptidase P and Y digestion

On-target car-boxypeptidase P and Y digestion

1. Place 1.0 µl matrix 2 onto the target.

   Note: The amount of matrix 2 must be optimized for each different target; it should dry in 10 s maximum (Vorm et al. 1994).

2. Dissolve the sample in 0.8 µl 0.1% trifluoroacetic acid and place it onto the dry matrix on the target.

   Note: A control peptide (e.g. 5 pmol substance P) should be used simultaneously to ensure the quality of the carboxypeptidases and to check the calibration of the mass spectrometer.

3. Wash the sample after drying by placing 2.0 µl water on the top of the sample and blowing away the water immediately (e.g., by pressurized air).

4. Introduce the dried target into the mass spectrometer.

5. Record the mass spectra.

   Note: If no signal appears repeat steps 3 and 4.

6. Export the target out of the mass spectrometer.

7. Place 0.8 µl carboxypeptidases P and Y (1:1) onto the sample on the target and let dry. The ratio of enzyme to substrate should be about 1:50 by weight.

8. Repeat step 3 twice.

   Note: If the mass spectra show many sodium adducts the washing procedure must be repeated again.

9. Introduce the dried target into the mass spectrometer.

10. Record the mass spectra.

11. Calculate the mass differences from the highest (the parent peptide) to the lowest mass. The mass differences allow determination of the cleaved amino acids from the carboxy-terminal (mass differences as above).

    Note: In many cases the sodium adducts are recorded in the mass spectra.

20.4
Results and Comments

The mass spectra of the time monitored carboxypeptidase P and Y-digest of substance P (1347.5 Da) are shown in Fig. 2. The masses of four truncated peptides were found after 1 min. (Fig. 2A). An additional mass was detected after 5 min carboxypeptidase incubation (Fig. 2B). No further degradation occurred within 1 h. The mass differences from the highest to the lowest mass were: 130.3 (Met), 113.1 (Leu/Ile), 57.2 (Gly), 147.1 (Phe) and 275.3 (Phe+Gln/Phe+Lys). Detection of the truncated peptide with the mass of 754 was very weak. In this case the mass difference of the degraded dipeptide allows determination of the released amino acids with unknown succession.

The mass spectra prior to (Fig. 3A) and after (Fig. 3B) on-target carboxypeptidase digestion of desmin 414–427 (1649.6 Da) allowed the carboxy-terminal sequence determination of three amino acids. The mass differences of the sodium adducts (+22 Da) from the highest to the lowest mass were 156.4 (Arg), 146.9 (Phe) and 114.2 (Asn). The desmin fragment was derived from a 2-DE-gel of human myocardial heart proteins (Jungblut 1994). The protein was blotted onto a PVDF membrane. After digestion with trypsin the peptides were eluted from the membrane and separated by microbore HPLC. The peptide mass fingerprint and the carboxy-terminal sequence information unequivocally identified the protein as desmin.

20.5
Troubleshooting

- The enzymatic release of different amino acids differs considerably. Sometimes no degradation occurs at all.

- Internal cleavage and release of dipeptides may occur.

- The first released amino acids may not be detectable if the degradation is too fast. In this case the experiment can be repeated with a new sample at a lower temperature (e.g., 10 °C). For time-monitoring experiments two additional aliquots should be taken: one immediately after addition of the carboxypeptidases and one after 30 s.

- The degradation of the amino acids is sometimes too slow. For time-monitoring digestion experiments the remaining sample (3.4 µl) can be incubated at higher temperature (e.g.,

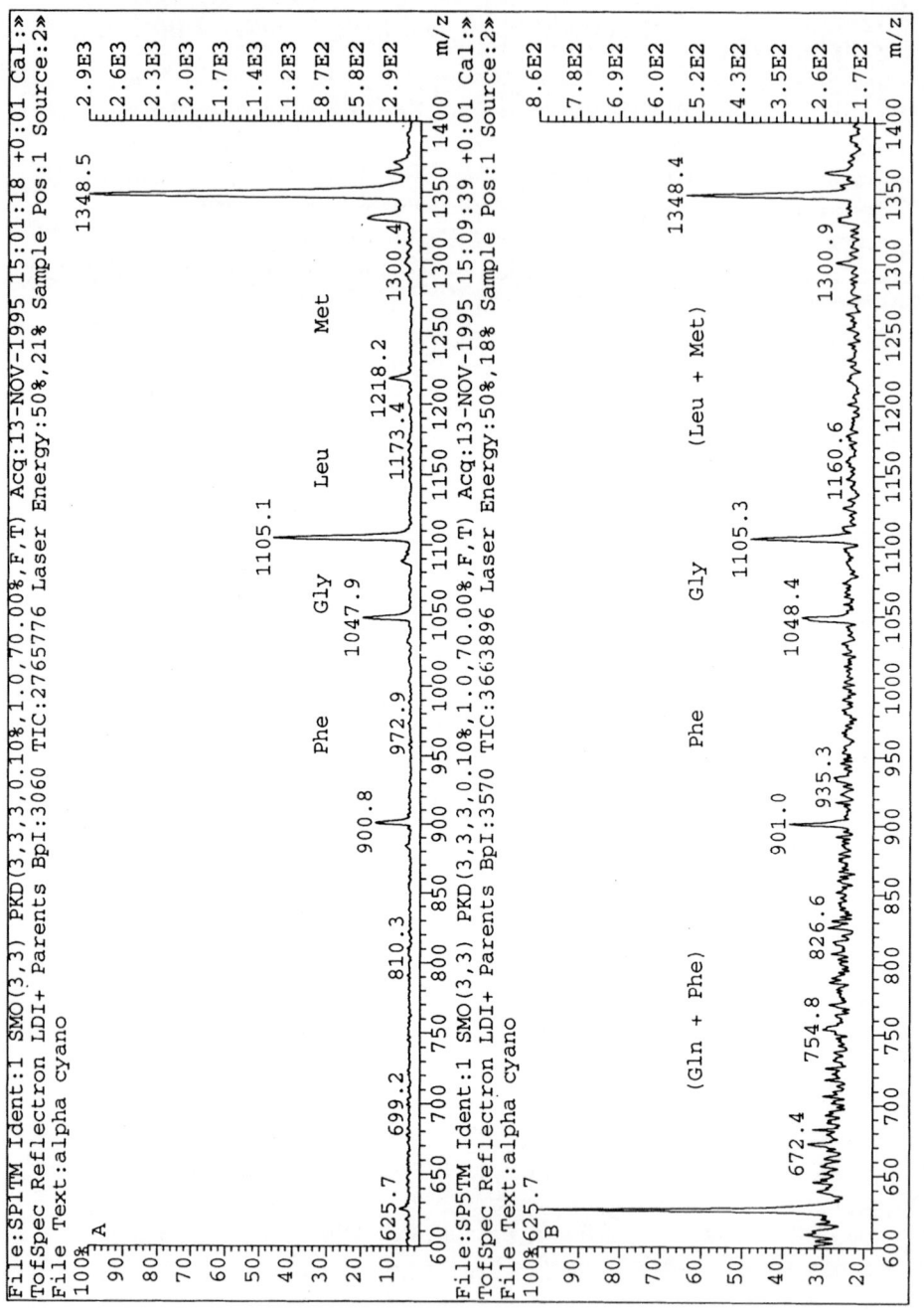

Fig. 2A, B. Mass spectra of the time-monitored carboxypeptidase P and Y digest of substance P after A 1 min and B 5 min

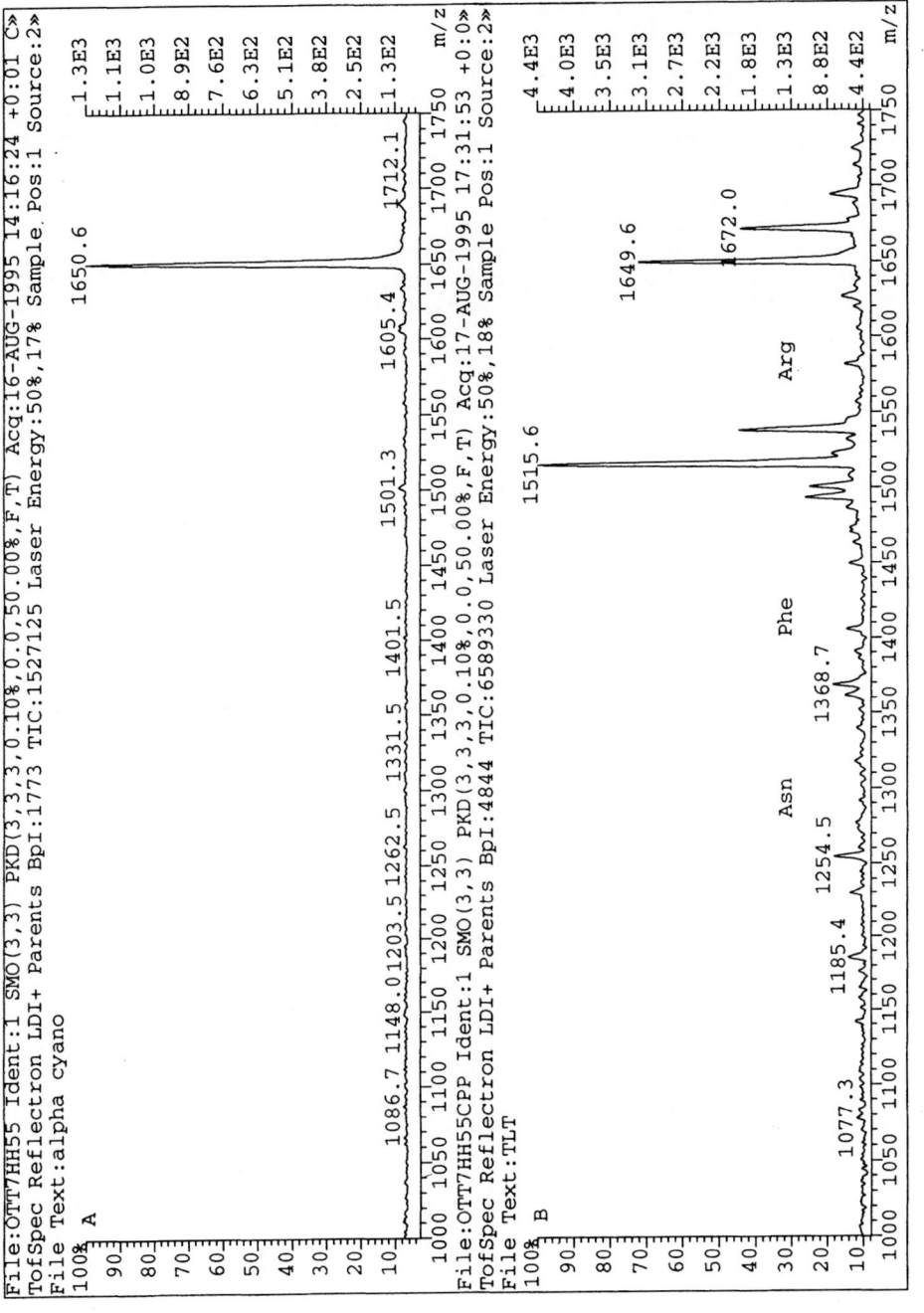

Fig. 3A, B. Mass spectra of the on-target carboxypeptidase P and Y digest of desmin 414–427 **A** before and **B** after digestion with carboxypeptidases P and Y

37 °C) for a longer time. For on-target degradation a longer incubation time can be obtained by applying more volume (e.g. 2 µl) of the carboxypeptidase solution.

References

Breddam K (1986) Serine carboxypeptidases. A review. Carlsberg Res Commun 51:83–128

Edman P, Begg G (1967) A protein sequenator. Eur J Biochem. 1:80–91

Henzel WJ, Billeci TM, Stults JT, Wong SC, Grimley C, Watanabe C (1993) Identifying proteins from two-dimensional gels by molecular mass searching of peptide fragments in protein sequence databases. Proc Natl Acad Sci 90:5011–5015

Hillenkamp F, Karas M, Beavis RC, Chait BT (1991) Matrix-assisted laser desorption/ionization mass spectrometry of biopolymers. Anal Chem 63:1193A-1203A

Inglis AS (1991) Chemical procedures for C-terminal sequencing of peptides and proteins. Anal Biochem 195:183–196

Jungblut P Otto A, Zeindl-Eberhart E, Pleißner KP, Regitz-Zagrosek V, Fleck E, Wittmann-Liebold B (1994) Protein composition of the human heart: The construction of a myocardial two-dimensional electrophoresis database. Electrophoresis 15:685–707

Pappin DJC, Hojrup P, Bleasby AJ (1993) Rapid identification of proteins by peptide-mass fingerprinting. Current Biology 3:327–332

Thiede B, Wittmann-Liebold B, Bienert M, Krause E (1995) MALDI-MS for C-terminal sequence determination of peptides and proteins degraded by carboxypeptidase Y and P. FEBS Lett 357:65–69

Vorm O, Roepstorff P, Mann M (1994) Improved resolution and very high sensitivity in MALDI TOF of matrix surfaces made by fast evaporation. Anal Chem 66:3281–3287

Chapter 21

Crystallization of Biological Macromolecules

F. FRANCESCHI

21.1
Introduction

To write about the crystallization conditions for a specific protein or nucleo-protein complex will be of no use to somebody
intending to crystallize a completely different molecule or even
an homologous one coming from a different organism. Therefore, the emphasis of this chapter is on the general aspects of
crystallization and on practical tips concerning the source of the
macromolecules, separation techniques, and crystallization
conditions.

It is generally believed that "protein crystal growth is an art
more than a science." One of the reasons for this belief can be
traced to the early days of structural biology, when researchers
were more concerned about the development of X-ray methods
than understanding the crystallization process. This concern
influenced the course of the research such that efforts to turn
protein crystal growth into a science are relatively recent (Feher
and Kam 1985).

Crystallographic studies of biomacromolecules have helped in
establishing modern biochemistry and molecular biology. These
studies have also revealed the relationships between structure
and function that are fundamental to understanding how enzymes, nucleic acids or their complexes with proteins operate in
biological systems. To be able to use this powerful method there
is an absolute requirement for good quality crystals, i.e., suitable
for structural analysis, and X-ray crystallography. However, obtaining high quality crystals remains problematic.

From the physical point of view, crystals are regular three-
dimensional arrays of molecules or molecular assemblies. An
ideal crystal can be seen as an infinite repetition of a perfect array, in which the asymmetric units (the building blocks of the
crystal) are arranged in a well-defined three-dimensional way.

However, the crystals obtained experimentally are finite and not perfectly ordered.

To date, the theoretical and practical aspects of crystallization of small molecules are mostly understood, and over the last decade serious efforts have been made to extend these studies to biological macromolecules and to rationalize the approaches used for their crystallization (Durbin and Feher 1990; Ihibault et al. 1992). Nevertheless, crystallization of biological macromolecules is still a process that is often nonreproducible and unpredictable. Indeed, the degree of irreproducibility and unpredictability seems to be directly proportional to the degree of complexity of the biological assembly.

21.2
Biological Macromolecules for Crystallization

In crystallization, it is of crucial importance to obtain a sufficient amount (normally a few milligrams) of highly purified biological macromolecules. However, the biomolecules subject to crystallization are normally present in minute quantities in wild-type cells. In a large number of cases this problem can be bypassed thanks to the advances of genetic engineering, which have made cloning and overexpression of genes in bacterial or eukaryotic cells possible, thus allowing us to obtain sufficient amounts of proteins (Bugg 1986). Despite careful consideration of the expression system to be used, some eukaryotic proteins are known to fold incorrectly or, even worse, to be insoluble when expressed in *E. coli*. In some cases this problem has been solved by mutating the protein in order to increase solubility (Schulze et al. 1994).

Source of the biological material During the last few years, it has become clear that the enzymes or ribonucleoprotein complexes isolated from "extremophile bacteria" are superior, in terms of stability, compared to their homologues isolated from mesophilic bacteria or eukaryotes. Extremophile bacteria are those that grow under extreme enviromental conditions such as high temperature or high salinity. For example, *Thermus thermophilus* has an optimal growth temperature of 75 °C and *Haloarcula marismortui*, a halophilic archaebacteria, has an intracellular potassium concentration of 4 M. These observations have prompted emphasis (with excellent results so far) on crystallizing proteins from extremophiles. Thus, *Bacillus stearothermophilus*, *Thermus thermophilus* and archaebacteria living under extreme environmental conditions are becoming frequent sources of macromolecules for structural studies, as can be seen in the significant increase in the number of crystals obtained from these organisms (reported in the Biological Macromolecule Crystallization Database, in the World Wide Web at "http://ibm4.carb.nist.gov:4400/bmcd/bmcd.html"; also see Gilliland. et al. 1994).

21.3
Purification and Handling of Biological Macromolecules

Purification, handling, stabilization and storage of macromolecules are essential steps before crystallization. Purification methods for biological macromolecules differ widely due to the different properties and stability ranges. The purification method of choice will depend mainly on the nature of the molecule itself; in most cases, the development and improvement of purification protocols is achieved by trial and error. Ideally, the purification steps prior to crystallization should involve no more than three to four chromatographic steps, because a partial loss of the activity due to the purification procedure results in increased microheterogeneity which results in a reduced tendency to crystallize (see below). Care should be taken to avoid the cofractionation of proteases or nucleases (Lorber and Giegé 1992). During chromatography, it is also important to avoid steps that could irreversibly unfold the protein; as a rule steps involving urea or other denaturing agents should be eluded. The techniques for purification will be also dictated by the source of the macromolecule to be crystallized. For example proteins from thermophilic organisms, when overproduced in *E.coli*, can be freed from the majority of the *E. coli* cytosolic proteins by heating at 60°–70 °C for 1 h; in this case the thermophilic protein remains in solution while the *E.coli* proteins tend to aggregate and precipitate (Desai and Pfaffle 1995).

The handling and storage of macromolecules must be done with particular care to avoid damage or loss of activity before crystallization. The macromolecules must therefore be kept in solvents having properties close to those of the cellular medium. In this respect the optimal buffer (i.e., Hepes, whose pK is little affected by temperature), pH, ionic strength, reducing agent (e.g., DTT, DTE, 2-mercaptoethanol, etc.) has to be found for each macromolecule.

Once the macromolecules have been purified, it is important to take into account that purity criteria for crystallization studies differ from the normal purity criteria. When one talks about "crystallography grade" purity, the molecules in question must not only be free of unrelated macromolecules or undesired small molecules, but they must also be as homogeneous as possible in terms of structural conformation. In other words, if good crystals of biological macromolecules are desired, the starting material to be crystallized must not only be pure but also a homogeneous population of a single conformer (Lorber and Giegé 1992), as microheterogeneity is one of the major causes for poorly diffracting crystals. Protein isoforms are normally caused by modifications such as phosphorylation, acetylation, and methylation, and these modifications have an effect on the isoelectric point. Although it is impossible to separate such isoforms by conventional chromatographic techniques, a chamber with isoelectric membranes (Hoffer Pharmacia Biotech Inc) that have user defined pH values (Weber et al. 1994) can be used in order to achieve separation. This procedure, called PrIME, allows the construction of membranes which differ by as little as 0.01 pH units and has been shown to improve crystal quality from isoforms of epidermal growth factor receptor (Weber et al. 1994).

21.4
Crystallization

As mentioned above, one of the fundamental prerequisites for solving the structure of a macromolecule is the existence of good-diffracting crystals. However crystals from biological macromolecules are especially difficult to grow, and most of them are rather small in size, making these crystals extremely fragile and very sensitive to the external environment.

When crystallizing macromolecules (protein, nucleic acids, ribonucleoprotein complexes or a virus) the formation of a supersaturated solution (a thermodynamically metastable state) is required. Once the crystallization starts the system moves to equilibrium. The size of the crystal will then depend on the degree of supersaturation and the kinetic pathway of nucleation and growth (Arakawa and Timasheff 1985). In general three distinct stages can be recognized: (1) formation of stable nuclei, (2) growth of nuclei to form crystals, and (3) termination of growth. An important point to recognize is that the conditions for most probable nucleation and crystal growth are not necesarily the same. We tend to assume that once microcrystals appear they should be left undisturbed to grow. However, the first appearance of a microcrystal could in fact be the sign to change to other conditions more favorable for growth (McPherson 1991).

Crystallization methods There are many methods to crystallize biological macromolecules. The most popular ones are based either on vapor-phase equilibrium or dialysis. Recent attempts to automate the crystallization process, or at least the screening for the right conditions, have mainly used microbatch crystallization methods (Chayen et al. 1990) which involve a commercial crystallization robot (Douglas Instruments, London, Great Britain).

The microcrystallization methods based on vapor diffusion, first used for the crystallization of tRNA, are currently the most widely used for the crystallization of biological macromolecules. One advantage of these methods (hanging drop, sitting drop; Fig. 1) is that they allow the use of very small volumes, i.e., small amounts of protein. In a typical experiment the droplet, containing the biomacromolecule to be crystallized (normally between 5 and 10 μl), an appropriate buffer, crystallizing agent(s) and other additives, is equilibrated against a reservoir (with a volume around 100 times bigger than that of the droplet). The reservoir contains the same solution as the droplet but without the biomolecules, and usually has twice the concentration of crystallizing agent. Equilibration proceeds by diffusion of the volatile species (water or organic solvent) until the vapor pressures in the droplet and the reservoir are equilibrated. This leads to a droplet volume change with a resulting increase of the concentration of the crystallization agent, hopefully leading to supersaturation and subsequent crystallization. Normally, these crystallization experiments are carried out in Limbro boxes, plastic boxes normally used for cell culture. Every plate has 24 wells, each with a volume of approximately 2 ml. In the hanging drop method, for example, the protein droplet is placed on a cover slip, the well-rims are greased with silicon grease, and the cover slip containing the droplet is inverted and used to close the well. In this case,

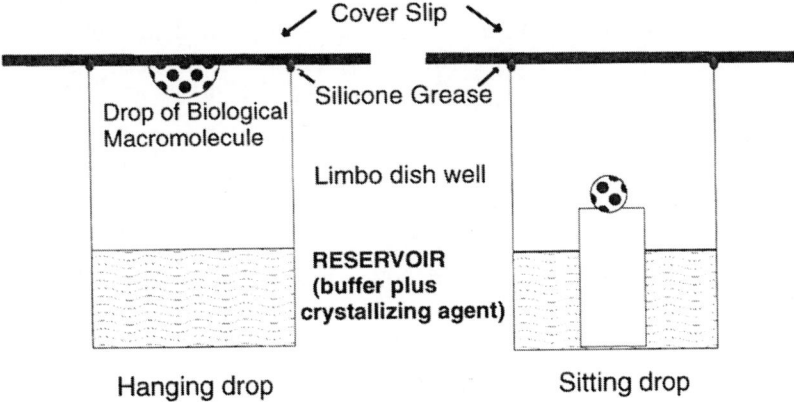

Fig. 1. The hanging and sitting drop crystallization method

the drops "hang" off the lid, and once the system is closed equilibration begins.

Crystallization agents   One of the crucial steps for any type of crystallization method, especially for those involving biomolecules, is to use the right crystallization agent. From the 1557 (100%) biological molecules reported to crystallize (Gilliland et al. 1994), the most frequently used crystallization agent by far is ammonium sulfate (40%), followed by polyethylene glycol (30%), phosphate salts (15%), organic solvents such as 2-methyl-2,4-pentanediol (10%) and low molecular weight alcohol (9%). In many cases more than one crystallization agent is simultaneously used.

Since the crystallization agents are normally used in relatively high concentrations during the crystallization process, it is of special importance to use commercially available high grade products. Alternatively these compounds can be repurified following well established techniques: recrystallization, in the case of ammonium sulfate or PEG; column chromatography; and either distillation or chromatography in the case of 2-methyl-2,4-pentanediol (Lorber and Giegé 1992).

Factors affecting crystallization   Some of the factors that influence the crystallization of macromolecules are: pH, temperature, ionic strength, concentration of the macromolecule and concentration of the precipitant. To this list, many other factors can be added, including: purity, additives, source of the macromolecules, contamination by bacteria or fungi, atmospheric pressure, metal ions, vibration, volume of the sample, reducing or oxidizing environment, electric or magnetic fields, handling by investigator, presence of dust, and detergents (Giegé and Ducruix 1992). It is clear that it is practically impossible to have absolute control over all the different factors. It is also clear that crystallization of a macromolecule which has not been crystallized before cannot be approached in a way that examines one factor at a time, but rather multifactorially, in which the effects of various conditions are simultaneously investigated.

Crystallization strategies   The "right" combination of factors that will produce crystals, especially of new macromolecules, is extremely hard to find a priori. One way to establish the conditions that will produce crystals is to use factorial design crystallization experiments. Factorial designs are important for screening large number of conditions in a comparatively low number of experiments. These experiments also provide information about the main effect of a number of possible factors and their interrelationship (Carter 1992). Another popular approach to tackling the crystallization of a new macromolecule is the "sparse matrix" proposed by Jancarik and Kim (1991), in which a set of crystallizing agents is tested using 48 different conditions. Although this matrix is not really a factorial approach it reflects the distribution in the use of different crystallization agents. In other words, the matrix is biased and selected from known crystallization conditions for macromolecules. For the initial screening of these conditions the 48 different solutions can be obtained commercialy (Hampton Research, Inc). It must be clear that this approach does not guarantee the production of crystals. Nevertheless, if crystals are not obtained the acquired solubility information can be used to refine the conditions around the regions where precipitated protein is obtained.

After good diffracting crystals have been grown, obtaining the phase information emerges as the next crucial problem. One of the most widely used methods to obtain the phase information (isomorphous replacement) requires cocrystalization of the macromolecules (or the soaking of existing crystals) with heavy atoms. Introducing heavy atoms in crystals that are isomorphous with respect to the "native" or unmodified ones is a prerequisite for structural studies. The conditions for the growth of crystals from the "derivatized" molecules can vary considerably with respect to the native crystals. In the case of soaking existing crystals in heavy atom solutions, conditions must be found for the soaking procedure as well as for handling the soaked crystals.

The search for derivatives

21.5
Some Advances in Data Collection Relevant to Crystallization

One important advancement in reducing the amount of material needed for structural studies is the use of cryotechniques for data collection. This approach dramatically reduces radiation damage of the crystals, thus allowing collection of complete data sets from single crystals (Hope et al. 1989). Another of the most frequent problems, namely, not having crystals large enough for structural analysis, has been at least partially solved by the development of high-brilliance synchrotron beamlines, in which crystals as small as 50µm can be effectively illuminated (Lattmann 1994).

21.6
Crystals of Large Macromolecular Assemblies

The study of crystals from large biomacromolecules, and especially those with very large unit cells, is particularly difficult. Due to the large number of atoms in these assemblies, there are more reflections to be measured in the diffraction patterns, and they are very closely spaced. The first very large structures to be crystallographically solved were the icosahedral viruses; however, in this case, the high symmetry of the particles provided some relief from the difficulties mentioned above (Lattmann 1994).

An extremely challenging structure that has been crystallized is the bacterial ribosome. A typical bacterial ribosome weighs around 2.3 million Daltons, contains around 60 different proteins and has three RNA chains of different lengths and composition. Unfortunately, the bacterial ribosome lacks any internal symmetry. Nevertheless, it has been possible to crystallize whole ribosomes (70S) and both of their subunits (30S and 50S). Crystals of the 50S subunits of *Haloarcula marismortui* diffract to almost atomic resolution, 2.9 Å (Berkovitch-Yellin et al. 1992; Franceschi et al. 1993). Recently, significant progress towards an intermediate resolution structure has been made by soaking the ribosomal crystals in solutions containing different tungsten clusters, and preliminary electron density maps at medium resolution have been obtained. (Franceschi et al. 1995).

Suppliers

Hampton Research 25431 Cabot Road, Suit 205 Laguna Hills CA 92653–5527 USA. Fax 1–714–6991040 Tel. 1–714 586 1453. E-mail: XTALROX@AOL:COM. Crystallization screening kits and most of the materials needed for crystallization

Hoefer Pharmacia Biotech Inc. 654 Minnesota Street. P.O. Box 77387 San Francisco, California 94107–3807. Tel 1–415–282 2307, Fax 1–415–821–1081. PrIME system for separation of protein isoforms

Flow Laboratories Inc. Mc Lean Virginia 22102 U.S.A Limbro plates

References

Arakawa T, Timasheff SN (1985) Theory of protein solubility. Methods in Enzymol 114: 49

Berkovitch-Yellin Z, Bennet WS, Yonath A (1992) Aspects in structural studies on ribosomes. CRC Rev Biochem Mol Biol 27:403

Bugg CE (1986) The future of protein crystal growth. J Cryst Growth 76:535

Carter CW Jr.(1992) Design of crystallization experiments and protocols. In: Ducruix, A and Giegé, R (eds) Crystallization of nucleic acids and proteins: a practical approach. Oxford Univ., p 47,

Chayen N, Stewart P, Maeder D, Blow D (1990) An automated system for microbatch protein crystallization and screening. J Appl Cryst 23:297–302

Desai UJ, Pfaffle PK (1995) Single-step purification of a thermostable DNA polymerase expressed in Escherichia coli. Biotechniques 19: 780–784

Durbin SD, Feher G (1990) Studies of crystal growth mechanisms of proteins by electron microscopy. J Mol Biol 212:763

Feher G, Kam Z (1985) Nucleation and growth of protein crystals: general principles and assays. Methods in Enzymol 114: 77

Franceschi F, Weinstein S, Evers U, et al. (1993) .Towards atomic resolution of prokaryotic ribosomes: crystallographic, genetic and biochemical studies. In: K. Nierhaus et al. (eds) The translational apparatus, Plenum, NY, p 397

Franceschi F, Weinstein S, Sagi I, et al. (1995) The combination of functional, genetic, microscopical and crystallographical studies led to initial phasing of data collected from ribosomal crystals at intermediate resolution. In: Sarma R, Sarma M (eds) Biological structure and dynamics, vol 2. Adenine, New York

Giegé R, Ducruix A (1992) An introduction to the crystallogenesis of biological macromolecules. In: Ducruix A, Giegé R (eds) Crystallization of nucleic acids and proteins: a practical approach. Oxford Univ., p 7

Gilliland GL, Tung, M, Blakeslee, D.M, Ladner J (1994). The biological macromolecule crystallization database, version 3.0: new features, data, and the nasa archive for protein crystal growth data. Acta Crystallogr D50: 408

Hope H, Frolow F, von Böhlen K, Makowski I, Kratky C, Halfon Y, Danz H, Webster P, Bartels K, Wittmann HG, Yonath A (1989) Cryocrystallography of ribosomal particles. Acta Crys. B45: 190

Ihibault I, Langowski J, Lebermann R (1992) Optimizing protein crystallization by aggregate size distribution analysis under dynamic light scattering. J Cryst Growth 122: 50

Jancarik J, Kim SH (1991) Sparse matrix sampling: a screening method for crystallization of proteins. J Appl Cryst 24:409–411

Lattman E E (1994) Protein crystallography for all. Proteins: 18:103

Lorber B, Giegé E. (1992) Preparation and handling of biological macromolecules for crystallization. In: Ducruix A, Giegé R (eds) Crystallization of nucleic acids and proteins: a practical approach. Oxford Univ., p. 19

McPherson A (1991) Useful principles for the crystallization of proteins. In: H Michel (ed) Crystallization of membrane proteins. CRC, Boca Raton, p 1

Schulze AJ, Degryse E, Speck D, Huber R, Bischoff R (1994) Expression of alpha 1-proteinase inhibitor in Escherichia coli: effects of single amino acid substitutions in the active site loop on aggregate formation. J Biotechnol 32: 231

Weber W, Wenisch E, Guenther N, Marnitz U, Betzel C, Righetti PG (1994) Protein microheterogeneity and crystal habits: The case of epidermal growth factor receptor isoforms as isolated in a multicompartment electrolyzer with isoelectric membranes. J Chromatog 679:181

Subject Index

AARE 173
Acetylserine 169
Acetylthreonine 169
Acid hydrolysis 107
Acylamino acid releasing enzyme 169
Acylation 269
Affinity chromatography 11
– antibody-antigen binding 12
– binding forces 12
– biomimetic interaction 14
– biovidin-avidin complex 19
– chaotropic salts 19
– detergents 19
– elution 19
– flow rate 18
– group specifity 13
– hydrophobic interaction 19
– immobilization 15
– immobilized metal ion affinity chromatograpy 14
– lectins 13
– macroporosity 15, 18
– metal chelating ligand 20
– mono-specific ligands 12
– pseudoaffinity chromatography 14, 22
– recombinant proteins 21
– site-directed immobilization 16
– space 6
– specific binding 17
– specific interaction 11

– specifity of binding 11
– synthetic polymers 15
– tentacles 15
Anhydrous hydrazine 171
Amino acids 249
Amido black 185
Amino acid analysis 185, 225, 231, 246
– advantage 246

– maintenance of columns 246
Amino acid sequencing 133
Ampholytes 184
AQC derivatization 234
Aqueous two-phase systems 31
– agitator bead mill 37

Bacillus cereus 43
– cell homogenate 35
– electrostatic potential 33
– extractive purification 39
– high pressure homogenizator 37
– leucine dehydrogenase 43
– liquid-liquid partition 31
– partition coefficients 31
– partition conditions 37
– periplasmic human-like growth factor 145
– phase diagrams 31
– scale-up 43
– S. cerevisiae 37
Biotin-avidin 177
Blocked proteins 167
– N-acetylated proteins 172
– N-formylated proteins 170
Blocking 168
Blot reactor 117
Blotting 75
– efficiency 215
– mechanism 215, 217
– PVDF membrane 75
– semi-dry 215
Buffer preservation 255

Calculation of amino acid composition 245
Carboxypeptidases 296
Carboxy-terminal sequencing 296
Cartridge 117
Chemical cleavage 61, 64
– cleavage at methionine 65
– cleavage procedures 64

CNBr-cleavage 65
– partial acid hydrolysis 67
– sample preparation 65
Cleavages of proteins 62
– chemical 61, 62
– enzymatic cleavages 62, 94
Coomassie brilliant blue 185, 220, 224
Continous cross-current extraction 45
– large scale protein recovery 45
Crystallization 303, 306
– biological macromolecules 304
– factors 308
– microcrystallization 307
– proteins 306
Cystein residues 259, 260
– covalent chromatography 260
– cystein containing peptides 262
– thiopropyl Sepharose 260, 261, 263

D-amino acid 249
DABITC 112, 143
Dabsyl chloride 234, 242
– derivatization 241
DABTH amino acids 112, 145, 150
Deblocking 168
– sequential deblocking 175
Derivatization 231, 252
Desalting of peptides 87
Digestion 75
– in situ 75, 63, 85, 92
Dinitrophenyl-(DNP)-peptide 107
Disulfide bonds 259, 261
– cystine 259

Edman degradation 107, 137 153
Electrophoresis 85, 283
– reagents 139
– solvents 139
Enzymatic cleavages 61, 62
– Arg-C 67
– Asp-N 69
– chymotrypsin 62, 68
– Glu-C 68
– in situ digestion 62
– limited proteolysis 62
– Lys-C 67, 90, 98
– protease concentration 63
– proteins immobilized 63
– proteolytic enzymes 62
– thermolysin 68
– trypsin 67, 90, 98
Extraction of peptides 88

Elektrospray ionization mass spectrometry 279
ESI-MS 279

Farnesylation 270
Fast red 223
Fatty acid 269
Flow-through reactor 117
Fluorescence detection 238, 243
FMOC 240
FMOC derivatization 233, 238

Genetic approach 133
Geranylgeranylation 270
Ghost peaks 247
Glass fiber membranes 215
Glycoproteins 289
Gradient HPLC 115

High performance liquid chromatography 113
– gradient 115, 124
– isocratic 115
HPLC 148
– DABTH 148
HPLC purification 74, 76
– amino acids 49, 54
– C4-, C8-, C18 alkylated support 77, 78
– cytochrome C tryptic peptides 49, 54
– eluents 49
– large biomolecules 49
– maintenance of columns 57
– microbore reversed phase 90, 98
– microseparation of peptides 86
– narrow-bore reversed phase 90, 98
– nonporous silica-based particles 49
– peptide mapping 74, 93
– peptide mixture 86
– proteins 49
– rechromatography 99
– RP-HPLC 75
– sample preparation 51
– troubleshooting 56

Immobilized metal chelate affinity chromatography 20
– angiotensin 25
– bradykinin 25
– neurotensin 25
– peptides 25
– samostatin 25
Immnoblotting 225

Immunostaining 185, 215
Inert gas 115
Internal standard 251
In-gel digestion 98
Initial yields 114
Internal peptides 61
– internal sequence 85, 97
– internal sequencing 225
Isoelectric focusing (IEF) 183
– cathodic drift 184, 210
– immobilized carrier ampho-
  lytes 184
Isoprenylation 273

Ladder sequencing 282
L-amino acids 249
L-homo-arginine 251
L-norvaline 251
Lectin staining 215

MALDI-M 94, 280, 282, 295
Manual sequence analysis 138
– extraction steps 150
– reagent stability 149
– troubleshooting 149
Mass mapping 286
Mass spectrometry 133, 134, 177,
  185, 225, 279
Matrix-assisted laser desorp-
  tion/ionization mass spectrometry 280, 295
Microsequencing 107, 141
– conversion 110
– cyclization 110
– preparation of sample 142
– subractive Edman 111
Microseparation of peptides 86
– micropurification 97
Modifications 133
Myristoyl group 174
Myristoylation 269

N-acetylated proteins 172
NBDCl 233
– precolumn derivatization 237
N-formylated proteins 170
NH$_2$-terminal sequencing 224
N-isobutyryl-L-cysteine 250
Nitrocellulose 215

On-line HPLC 123
OPA 235, 244, 249
OPA derivatization 233, 235
OPA postcolumn derivatization 242

Overlap 113

Palmitoylation269
Phenylisothiocyanate (PITC) 108
Phenylthiocarbamoyl peptide 110
Phosphorylation 286
Phosphoproteins 287
Poinceau S 185
Polybrene 116
Polyvinyledenedifluoride 185
Postcolumn derivatization 234, 244
Posttranslational modification 269
Precolumn derivatization 231
Prenylation 273
Protein modifications 61
Protein sequence database 108
Protein sequencing 282
PTH amino acid derivatives 111
– identification 117
– standards 124
Purification of proteins 3, 31
– aqueous two-phase 31
– chaotropic reagents 5
– fusion proteins 6
– inclusion bodies 5
– membrane proteins 6
– recombinant proteins 5
PVDF membranes 168, 215
Pyroglutamic acid 171
Pyroglutamyl peptidase 171

Recombinant proteins 5, 21, 45
Refolding of proteins 5
removal of cell debries 34
Repetitive degradation yields 113,
  114

Semi-dry blotting 215
Sequenator 108
– cleavage 122
– conversion 122
– coupling 121
– degradation programs 126
– filter washes 125
– gas-phase delivery 119
– hybrid radiator valve system 120
– liquid-phase delivery 119
– on-line detection 122
– reagents 121
– sequencer program 118
– solvents 121
Sequence-tag 286
Sequential deblocking 175
Separation of amino acids 49

Separation of peptides   73, 49, 52, 91
– HPLC   73, 74, 75, 49
– Thin-layer fingerprints   73, 74, 78
Separation of proteins   49, 52
Sodium dodecylsufate polyacrylamide
    electrophoresis   183
Solid phase sequencer   116
Solid phase sequencing   153
– aminopropyl glass   156
– aminophenyl glass   162
– coupling of peptides   163
– DITC coupling assay   160
– glass matrices   158
– immobilization with DITC   154
– immobilization with EDC   154
– lacton immobilization   156
– membrane matrices   158
– PVDF   157
Sulforhodamine   185, 221, 225

Tandem mass spectrometry   281
Thin-layer fingerprints   73
– ascending chromatography   74
– buffers and solutions   79
– electrophoresis   74
– elution of peptides   80
– first dimension chromatogra-
    phy   79

– fluorecamine   81
– ninhydrin   81
– second dimension chromatogra-
    phy   80
Thiophilic adsorption chromatogra-
    phy   22
– antibody purification   23
– human γ-globulin   28
– human serum albumin   28
– protein A column   23
– protein G column   23
– thiophilic regions   22
Trimethylamine   115
Two-dimensional electrophore-
    sis   184
– calibration of molecular mass   208
– calibration of pI   207
– databases   187
– detection   187
– matching   187
– resolution   184

Vinylpyridine   100

World wibe web   187